National Wildlife Refuge System

NATIONAL WILDLIFE REFUGE SYSTEM

DAVID CALLIHAN

Nova Science Publishers, Inc.
New York

For permission to use material from this book please contact us:
Telephone 631-231-7269; Fax 631-231-8175
Web Site: http://www.novapublishers.com

LIBRARY OF CONGRESS CATALOGING-IN-PUBLICATION DATA

ISBN: 978-1-60692-792-2

Available upon request

Published by Nova Science Publishers, Inc. ✝ New York

CONTENTS

PREFACE*

The US National Wildlife Refuge System was created by Executive Order On March 14, 1903, when President Theodore Roosevelt established the country's first wildlife refuge on Florida's central Atlantic coast C the Pelican Island National Wildlife Refuge (NWR). From its modest beginning on Pelican Island the Refuge System has expanded into a network of over 550 distinct units that encompasses over 95 million acres. Alaska contains approximately 76.8 million acres of refuge lands, or about 80% of the land in the total system. To accomplish its mission the Refuge System finalized a strategic plan in early 2007 that contains twelve strategic outcome goals (SOGs). These goals cover the areas of habitat and wildlife conservation, wildlife dependent recreation, law 1 National Wildlife Refuge System Improvement Act of 1997; Public Law 105 57 enforcement, fire management, welcoming and orienting visitors, wilderness management, conservation planning, infrastructure and equipment maintenance, strategic growth and organizational excellence. The Refuge System is part of the United States Fish and Wildlife Service, which is managed by the Department of Interior.

* This is an edited, excerpted and augmented edition of a U.S. Fisheries and Wildfife Service (U.S. Department of·the Interior) Summary Overview dated June 2008.

Acronyms

ABC	Activity Based Costing
AFWA	Association of Fish and Wildlife Agencies
AHWP	Annual Habitat Work Plan
API	Asset Priority Index
BLM	Bureau of Land Management
CCP	Comprehensive Conservation Plan
CNO	California/Nevada Operations
DOI	Department of the Interior
DU	Ducks Unlimited
EE	Environmental Education
ESA	Endangered Species Act
FCI	Facility Condition Index
FMP	Fire Management Plan
FWS	Fish and Wildlife Service
GSA	General Services Administration
HAPET	Habitat and Population Evaluation Team
HMP	Habitat Management Plan
IACP	International Association of Chiefs of Police
IPA	Interagency Personnel Agreement
MSI	Management Systems International
NEPA	National Environmental Policy Act
NFWF	National Fish and Wildlife Foundation
NGO	Non-Governmental Organization
NOAA	National Oceanic and Atmospheric Administration
NPS	National Park Service
NWR	National Wildlife Refuge
NWRA	National Wildlife Refuge Association

NWRS	National Wildlife Refuge System
OMB	Office of Management and Budget
RAPP	Refuge Annual Performance Plan
RIA	Refuge Improvement Act
RLGIS	Refuge Lands Geographic Information Systems
SAMMS	Service Asset Managment and Maintenance System
SHCI	Strategic Habitat Conservation Initiative
SOG	Strategic Outcome Goal
TNC	The Nature Conservancy
TPL	The Trust for Public Lands
VSP	Visitor Services Plan
WMA	Wildlife Management Areas
WMD	Wetland Mangement District
WUI	Wildland Urban Interfac

EXECUTIVE SUMMARY

The mission of the National Wildlife Refuge System is to administer a national network of lands and waters for the conservation, management and where appropriate, restoration of the fish, wildlife and plant resources and their habitats within the United States for the benefit of present and future generations of Americans [1]

This book is a summary of the complete evaluation and includes an overall performance rating, conclusions and recommendations for each of the Refuge System's twelve strategic outcome goals. A complete set of evaluation findings can be found in the full report - *An Independent Evaluation of the Effectiveness of the U.S. Fish and Wildlife Service's National Wildlife Refuge System* (MSI, June 2008). In addition to findings, the complete evaluation report contains response data from three surveys (refuge managers, state fish and game agencies, and Friends Groups/partners), a list of all persons interviewed approximately [1] 250), references, and a bibliography.

BACKGROUND

The US National Wildlife Refuge System was created by Executive Order On March 14, 1903, when President Theodore Roosevelt established the country's first wildlife refuge on Florida's central Atlantic coast – the Pelican Island National Wildlife Refuge (NWR). From its modest beginning on Pelican Island the Refuge System has expanded into a network of over 550 distinct units that encompasses over 95 million acres. Alaska contains approximately 76.8 million acres of refuge lands, or about 80% of the land in the total system.

To accomplish its mission the Refuge System finalized a strategic plan in early 2007 that contains twelve strategic outcome goals (SOGs). These goals cover the areas of habitat and wildlife conservation, wildlife-dependent recreation, law enforcement, fire management, welcoming and orienting visitors, wilderness management, conservation planning, infrastructure and equipment maintenance, strategic growth and organizational excellence. The Refuge System is part of the United States Fish and Wildlife Service, which is managed by the Department of Interior. The National Wildlife Refuge System collaborates closely with other Fish and Wildlife Service programs, including the migratory birds, endangered species and fisheries programs.

This book reviews the Refuge System's twelve strategic outcome goals and provides an assessment as to how well the system is doing in accomplishing each goal. In addition, a section on the Refuge System's operating context, which analyzes budget and administrative trends over the past several years, has also been included.

Evaluation Methodology

This evaluation was conducted between October 2006 and September 2007 and used a multi- method and multi-source data collection methodology [2]. MSI used a multi-source methodology to overcome the limitation of having to base analysis on a single source of information as single-source data may have weaknesses or unduly bias conclusions. In addition, a multi-method approach allows for a greater depth of understanding of particular issues.

The principal data collection processes used in this evaluation included:

Fish and Wildlife Service (FWS) and Partner Interviews: More than 250 interviews were conducted as part of this evaluation. Those interviewed included: Refuge System managers in Washington; a wide range of FWS field staff, from both regional offices and field stations; Refuge System stakeholders in Congress, the Department of the Interior and the Office of Management and Budget (OMB); national and local NGO partners; and staff from state fish and wildlife agencies.

Site Visits: The evaluation team visited all eight Refuge System regional offices and at least two refuges in each region. Refuge and regional office visits included meetings with refuge and regional office staff, state fish and game officials and non governmental partners, such as representative from the Audubon Association, Ducks Unlimited, the Nature Conservancy, Friends Groups and other local partners.

Refuge Managers Survey: An online Refuge Manager's Survey was conducted between March 21st and April 19, 2007. The survey was a combination of close-ended and open-ended questions structured to collect information on the implementation and effectiveness of the Refuge System's twelve strategic outcome goals. A total of 312 refuge managers completed the survey, which represents a survey completion rate of over 90% of current refuge managers.

Partners and State Fish and Game Surveys: Two additional surveys were conducted to solicit the views of Refuge System partners on the quality of their partnerships with the Refuge System and on their views of the Refuge System's effectiveness. These surveys were:

- A survey of local Partners and Friends Groups was undertaken from March 17-25, 2008. A total of 83 responses were received from 98 potential respondents. The response rate was 85%.
- A survey of officials from state fish and game agencies was conducted from April 29–May 16, 2008. Responses were limited to one response per state agency. A total of 32 states responded to the survey, which constitutes a response rate of 64%.

Review of Existing Data: This included documents and databases, both from the Refuge System and from other land management agencies – the latter which provided context and benchmarking. Analysis included a careful review of the Refuge System's annual performance monitoring database – the Refuge Annual Performance Plan (RAPP).

PRINCIPAL FINDINGS AND CONCLUSIONS

An overview of the performance of each Strategic Outcome Goal is provided below.

Operating Context: Refuge System budgets have been in decline over the past several years and actual purchasing power has declined about 11% (between the FY 2003 and the requested FY 2008 budget). During this period of budget decline some Refuge System costs have increased due to inflation and annual adjustments, e.g. salaries. As a result, the Refuge System has not been able to maintain its level of operational activity from one year to the next – services and personnel have had to be cut back.

Concurrent with declining budgets, the Refuge System has also experienced an increase in administrative requirements. Together, these factors have had a negative effect on the Refuge System's ability to achieve its core goals – refuge managers have less time, and less money, to focus on the accomplishment of their mission than was the case five years ago. The areas most impacted have included:

the Refuge System's ability to conduct adequate monitoring and inventory work; the law enforcement program, which simply has too few officers to enable the Refuge System to provide adequate law enforcement coverage; and the rate of growth of the Refuge System, which has declined markedly over the past five years.

SOG 1: Conserve Manage, and Where Appropriate, Restore Fish, Wildlife and Plant Resources and Their Habitats

Partially Effective: This objective is rated "Partially Effective" because of the significant amount of refuge land that is need of additional management attention and the inconsistent application of science-based management across the Refuge System. As per the Refuge System's RAPP performance reporting system, 89% of refuge lands – 76.5 million acres – are in Class I condition, which means the land is receiving needed management action or does not require additional management action at this time. Alaska's sixteen refuges report that 98% of their habitats were in Class 1 condition in 2006 (as per RAPP data reporting). However, for NWRS lands outside of Alaska, 59% of the 18.9 million acres were reported as being in Class 1 condition in 2006– meaning that 41% are in need of management attention.

A significant portion of refuges have not developed Habitat Management Plans and there is an insufficient level of biological inventory and monitoring work being done – only 11% of refuge managers surveyed described the current level of inventory and monitoring work as being mostly or fully sufficient.

SOG 2: Provide Quality Environments with Adequate Water

Unable to Evaluate: This objective is rated "Unable to Evaluate" as a result of the limited information available against which to undertake an assessment of this strategic goal. The Refuge System does not currently operate a well defined and structured water resources program. There is currently no individual or office designated to coordinate the Refuge System's water rights and water quality activities.

SOG 3: Ensure That Unique Values of Wilderness, Other Special Designation Areas, and Cultural Resources Are Protected

Partially Effective: The NWRS contains about 20.7 million acres of wilderness, of which approximately 90%, or 18.6 million acres, is in Alaska. In addition, about 1.9 million of proposed acres of wilderness exist in the NWRS. The NWRS currently operates under the 1986 Wilderness Stewardship Policy. This policy is outdated and does not provide Refuge Managers adequate guidance regarding permissible management actions. A new draft policy has been developed and was released for public comment in 2001 but has never been finalized. The NWRS has supported the development of wilderness training courses and refuge managers overwhelmingly feel these courses have been effective in enabling them to acquire the skills necessary to manage wilderness areas; 64% of refuge managers who manage wilderness areas have completed the required wilderness training.

SOG 4: Welcome and Orient Visitors

Partially Effective: The NWRS is reasonably effective in terms of informing and engaging refuge visitors but could easily improve its performance in this area. Brochures are generally informative and available at refuges, and refuge employees and volunteers are able to provide helpful and informative answers to visitor questions. However, videos and CDs – very engaging and effective means of providing information to refuge visitors - are substantially underutilized. The information provided on refuge websites is very inconsistent from refuge to refuge and frequently provides only the most basic information. The NWRS could do a substantially better job at orienting visitors by improving its websites and making sure website content is updated and consistent.

SOG 5: Provide Quality Wildlife-Dependent Recreation and Education Opportunities

Effective: The Refuge System has done a good job at expanding the number of refuges that offer wildlife-dependent recreation opportunities and, overall, the visitor satisfaction rate at refuges appears to be very high –above 90% in the 2002 and 2004 surveys (note: the surveys were conducted only at fifty high visitation refuges). In terms of the individual Big 6 recreational activities, the operation of hunting, fishing, wildlife viewing, and photography programs are generally operating at a satisfactory level in terms of the Refuge System's ability to provide an adequate level of service and in terms of the support provided to those programs by the Refuge System. The environmental education and interpretive programs, on the other hand, are not able to meet public demand and are not adequately supported by the Refuge System. This latter conclusion is based solely on the view of refuge managers: 55% of refuge managers surveyed indicated they are not able to adequately meet the demand for environmental education services and 48% indicated they are not able to meet the demand for interpretive services.

SOG 6: Facilitate Partnerships and Cooperative Projects to Engage Other Conservation Agencies, Volunteers, Friends, and Partners in the NWRS Mission

Highly Effective: This objective was rated highly effective for several reasons: over the past ten years the Refuge System has been able to significantly expand participation by volunteers and Friends Groups; partnerships with thousands of local and national organizations make a significant contribution to the accomplishment of the Refuge System's key objectives, particularly in the areas of habitat restoration and visitor services; and partnerships bring a tremendous amount of funding into the system – in 2005 alone the total value of partnership contributions to the Refuge System exceeded $50 million, with over $30 million of the total being in direct cash contributions.

State Fish and Game Agencies: 88% of state agencies rated the quality of their relationship with individual refuges as between good and excellent; 47% rated the quality of the relationship as excellent or very good.

Partner Agencies: 93% of partners rated the quality of their relationship with individual refuges as between good and excellent; 56% rated the quality of the relationship as excellent.

SOG 7: Protect Resources and Visitors through Law Enforcement

Ineffective: Low staffing levels are leading to a substantial and critical lack of law enforcement coverage and capability at many refuges across the system. At many refuges, law enforcement coverage is insufficient to ensure the protection of resources and the safety of visitors and refuge staff. A substantial majority of refuge managers (over 70%) feel visitor safety and law enforcement performance has declined in recent years. The issue of public safety is of particular concern given that only seven of the refuge managers from 50 high visitation refuges (with annual visitation in excess of 250,000) who responded to the MSI survey indicated that law enforcement coverage is sufficient on their refuge. It is highly unlikely that any meaningful progress towards improving the Refuge System's law enforcement capability can be achieved under current and expected budget allocation levels.

SOG 8: Provide Infrastructure and Equipment Adequate to Support Mission and Maintained in Good Condition

Effective: The most important refuge assets -- those most necessary to the achievement of refuge conservation and public use objectives -- are generally well maintained. Seventy-five percent of refuge managers surveyed feel that the assets most critical to their refuge's mission and purpose, such as water management systems, are maintained in a condition adequate to support and achieve those goals. An important caveat to this conclusion is the fact that a substantial minority of refuge managers (40%) believe their refuges require new facilities if they are to meet their purpose and objectives. In the mid-1990s, the maintenance of the Refuge System's infrastructure and equipment was a critical concern and the maintenance budget subsequently increased dramatically -- from $21 in 1996 million to $91.5 million in 2004 (a 336% increase over eight years in actual funding dollars). The availability of increased funds over the past seven or eight years has allowed the Refuge System to effectively address preventive maintenance requirements. Subsequent to 2004, however, maintenance funding dipped substantially – a decline of 30% from 2004 to 2007. It is important to note that if the recent backsliding in maintenance funding is not reversed infrastructure maintenance will soon once again become a critical problem SOG 9: Complete Quality and Useful Comprehensive Conservation Plans (CCPs) on Schedule and with Full Engagement of Partners.

Effective: The NWRS is required to complete CCPs for 554 refuges by 2012. To date, two hundred and five refuges have completed CCPs– or about 37% of required units (analysis as per mid-2207). Although the pace of CCP completion has accelerated significantly over the past few years, the Refuge System is slightly behind schedule in terms of meeting its CCP completion target. In April 2007, the Refuge System began implementing the 2012 Plan, an Action Plan to Meet Our Legislative Mandate, which lays out a series of actions intended to ensure that all required CCPs are completed by 2012. Overall, refuge managers have found CCPs to be a useful tool for clarifying objectives, guiding habitat management decisions, and clarifying public use decisions.

As per the MSI State Fish and Game Agency Survey: 94% of state agencies agreed or strongly agreed that they had been provided an opportunity to meaningfully participate in the CCP process; 95% of state agencies agreed or strongly agreed that their participation in the CCP process had improved their communication and coordination with the Refuge System.

SOG 10: Strategically Grow the System

Ineffective: This objective was rated ineffective for a number of reasons, including: the rate at which land has been added to the NWRS has declined significantly over the past five years; land purchased by the Refuge System often does not match the priorities identified by the NWRS' Land Acquisition Priority System, especially over the past few years; and the current DOImanaged land appraisal process that the NWRS uses is ineffective and cannot be relied upon to produce timely or accurate appraisals, resulting in available land deals being lost.

SOG 11: Reduce Wildfire Risks and Improve Habitats

Effective: This objective is rated "Effective" as a result of the systematic planning and execution by which the NWRS utilizes prescribed fire to improve wildlife habitat and reduce fuels loads and also for the Refuge System's ability to fight and suppress wildfires. Where refuges have the qualified staff and budget, the high level of planning, training, and coordination results in application of prescription fire to improve and maintain habitats, reduce fuel loads, and suppress unwanted wildfire. Based on MSI surveys and interviews, it appears that approximately one-half of the NWRS has the resources it needs – both budget and personnel – to use fire as a habitat management tool. For other units, issues of staffing, available budget, the growing percentage of Wildland Urban Interface lands, and the location of refuges relative to other fire resources impair the system's ability to promote prescription fire while proactively addressing fuels availability and effective wildfire suppression.

SOG 12: Promote and Enhance Organizational Excellence

Partially Effective: The Refuge System has introduced a number of new management and planning systems over the past several years, including a medium-term strategic plan, activity- based costing, RAPP work planning and reporting systems, and refuge-level comprehensive conservation planning. The Refuge System is also currently undertaking a Workforce Planning exercise to help better balance personnel and operational expenditures and to prioritize staffing and programs in consideration of declining budgets. The RAPP system has enabled the NWRS to better track and report on national- level accomplishments and the budget rebalancing exercise will, overtime, provide managers greater flexibility to address local priorities. The RAPP system, however, has not proved useful to analyzing program effectiveness nor is it used for program decision-making. In addition, there is significant inconsistency within the Refuge System in how policies and programs are implemented across regions. In particular, there is a great deal of variance in basic business management practices, such as budgeting, annual work planning and the use of station reviews/evaluations

PRINCIPAL RECOMMENDATIONS

This chapter contains specific recommendations for improving the effectiveness of the Refuge System in each of the individual strategic outcome goal sections. A brief summary of some of the recommendations most likely to improve performance are presented below.

The Law Enforcement program needs increased funding: There is a severe shortage of full-time law enforcement officers that can only be addressed by hiring additional full-time law enforcement officers—moving from current levels of around 200 full-time officers to at least 400 full-time officers. Implementation of this recommendation will require substantial resources, but an acceptable improvement in law enforcement coverage is of fundamental importance to the on-going effectiveness of the Refuge System.

The Refuge System should find a way to increase policy and program consistency across regions and between refuges: Part of this process could include standardizing budget development, work planning, reporting and evaluation requirements. Another aspect of this recommendation is the need to develop a clear point of authority and process for ensuring greater policy consistency.

Reduce administrative and reporting requirements: The Refuge System should strive to reduce administrative and reporting requirements - - particularly for smaller refuges (seven or fewer staff).

Hire Additional Biologists: As noted in the conclusion section, in part, the Refuge System is unable to fulfill its commitment to manage refuges using an adaptive management process because of a shortage of biologists (approximately 20% of the Refuge System's workforce are biologists). It is recommended that the Refuge System review the adequacy of its biology workforce as compared to system needs. The White Paper produced for the Conservation in Action Summit recommended that biological teams be added to the top 50 refuges. An assessment should be undertaken to determine the degree to which this has happened.

Biological monitoring and inventory work needs to be increased and a more consistent approach should be developed and implemented: The effort of developing a system-wide geographic monitoring capability should be continued and provided increased emphasis. For example, efforts should be made to build upon the Refuge Lands Geographic Information system (RLGIS) and accelerate and adequately resource the implementation of the Strategic Habitat Conservation Initiative. The Refuge System's challenge is to better define high- priority system-wide needs, identify best practices that meet these needs, and replicate these systems in an increased number of locations.

Develop a water strategy: The Refuge System should develop an overall strategy and management structure to more effectively assess and address water management issues. Steps to develop such a program would include appointing a Water Resources Coordinator and developing a policy, or at least a defined process, for how refuges should assess and manage water rights. As the Refuge System reviews the need to bolster its approach to water management it could also take the opportunity to review other program and issue areas that may benefit from increased attention, such as the impacts of climate change and its influence on how the Refuge System should be managed.

Develop consistent and improved refuge websites: Develop a single website format/architecture for each of the refuge unit's websites. There are several options available in terms of the approach used to manage refuge websites; however, the most efficient option would likely be to centralize the function in a single office or under a single contract.

Prioritize visitor services: In light of high public demand for wildlife-dependent recreation and the Refuge System's limited and stretched budgets, the Refuge System should prioritize the public use services it will offer and provide some guidance to refuges and regions as to how limited resources should be allocated among the various wildlife- dependent recreational activities. Particular attention should be given to better defining and supporting environmental education and interpretation programs, where such programs are appropriate and of high utility.

Strengthen the Refuge System's strategic growth program: The Refuge System should develop a Land Acquisition Policy and a corresponding strategy to guide expansion of the system. It is recommended that the land acquisition policy/system be developed to be consistent with the Fish and Wildlife Service's Strategic Conservation Habitat Initiative, a geospatial system being adopted by the Service to identify and monitor conservation priorities. In addition, the Refuge System should engage in a discussion with the Department of Interior to enable it to improve the process it uses to appraise potential real estate transactions as the current Department of Interior mandated Appraisal Services Directorate (ASD) system is ineffective.

Redesign the RAPP reporting system: This system should be redesigned based on a clarification of its purpose. If the system is to remain primarily an external reporting tool – for reporting to FWS, DOI, OMB, and Congress—then the system should be substantially simplified to focus on areas of key interest and the number of indicators tracked should be significantly reduced (by at least 50-60%). However, the RAPP system would be most useful to the Refuge System if it were redesigned to provide information that could help inform strategy and management decisions, which is not currently the case. This will require revising the system and also instituting practices to review and analyze the data for management decisions, e.g. an annual strategy and performance review workshop.

It is also recommended that the Refuge System disaggregate reporting data between the Alaska region and the rest of the system. Because approximately 80% of all refuge land is in Alaska, and more than 90% of this land is classified as wilderness, aggregating Alaska performance data with that from the rest of the Refuge System provides a distorted picture of the overall system's condition, needs and performance accomplishments

Develop a knowledge management program: The NWRS should implement a Knowledge Management System to foster information sharing, promote learning and to ensure that best practices are more widely disseminated and adopted. Consideration should be given to creating a dedicated Knowledge Management Unit, which would be responsible for program reporting (RAPP), archiving documents, managing evaluations, disseminating lessons and best practices, and responding to external information requests (together with public relations staff). The Unit's purpose would be to improve performance analysis and reporting and to raise the quality level of implementation practices across the Refuge System

Chapter I

INTRODUCTION

The National Wildlife Refuge System (NWRS) contains 96 million acres of many of the nation's most important conservation landscapes and is characterized by its proponents as "the most biologically diverse lands in America" [3]. The system contains representative landscapes of virtually all of the country's natural ecosystems and is critical to the health and survival of many migratory birds, endangered species, fish and resident wildlife. In addition, the Refuge System annually hosts over 34 million visitors, who engage in hunting, fishing and wildlife viewing, which makes the Refuge System one of the country's premier assets for supporting wildlife-dependent recreation. The system also serves as an important educational resource, as it annually provides over 800,000 environmental education opportunities to school children.

The Refuge System is part of the United States Fish and Wildlife Service, which is managed by the Department of Interior. The National Wildlife Refuge System collaborates closely with other Fish and Wildlife Service programs, including the migratory birds, endangered species and fisheries programs.

The National Wildlife Refuge System contracted with Management Systems International in September 2006 to conduct an independent evaluation of the overall effectiveness of its program. The evaluation's purpose was to identify program strengths, weaknesses, and performance information gaps, and to determine whether and to what degree the Refuge System is achieving its conservation mission. The evaluation was also conducted to help the Refuge System meet compliance requirements of the President's Management Agenda, including compliance with the Office of Management and Budget's Program Assessment Rating Tooll (PART) requirements.

This chapter is a summary of the complete evaluation and includes an overall performance rating, conclusions and recommendations for each of the Refuge

System's twelve strategic outcome goals. A complete set of evaluation findings can be found in the full report -An Independent Evaluation of the Effectiveness of the U.S. Fish and Wildlife Service's National Wildlife Refuge System (MSI, June 2008). In addition to findings, the complete evaluation report contains response data from three surveys (refuge managers, state fish and game agencies, and Friends Groups/partners), a list of all persons interviewed (approximately 250), references, and a bibliography.

EVALUATION OVERVIEW AND METHODOLOGY

This evaluation's design was built on a multi- method and multi-source methodological process of data collection. MSI used a multi-source methodology to overcome the limitation of having to base analysis on a single source of information; single-source data may have weaknesses or unduly bias conclusions. In addition, a multi-method approach allows for a greater depth of understanding of particular issues. For example, MSI's Refuge Manager Survey may highlight the strength of workforce viewpoints on any given issue, but interviews are required in order to understand the complexity of underlying issues and the reasons for particular ratings.

The principal sources/processes for data collection used in this evaluation have included:

FWS and Partner Interviews: Interviews have been conducted with senior national-level NGO staff, senior NWRS staff, the directors of the Migratory Bird and Fisheries Programs, and members of Congressional appropriations committees. All FWS Refuge Division Chiefs were interviewed at least once. In total, nearly 250 interviews were conducted as part of this evaluation.

- *Site Visits to Regional Offices*: The evaluation team visited all eight of the Refuge System's regional offices, where meetings were held with regional directors, refuge chiefs, refuge supervisors, planners, and law enforcement supervisors, among others.
- *FWS Staff Survey*: An on-line Refuge Manager's Survey was conducted between March 21 and April 19, 2007. The survey was a combination of close-ended and open- ended questions and was structured to collect information on the implementation and effectiveness of the Refuge System's twelve strategic outcome goals. A survey pre-test was conducted with NWRS managers and division chiefs prior to the survey's

release. A total of 312 refuge managers completed the survey, which represents a completion rate of over 90%.

- *Refuge Site Visits*: The following is a list of refuges visited during this evaluation. Site visits included visits to a minimum of two refuges in each FWS region. (See Table 1 below).

Table 1.

Refuge/Office	Dates
Minneapolis RO, Minnesota Valley, Fergus Falls	December 4–8, 2006
Denver RO, Rocky Mountain Arsenal	December 6 & 8, 2006
Charles M Russell NWR	January 11 & 12, 2007
Atlanta RO, Eufaula, Okefenokee	January 16–24, 2007
Mason Neck/Potomac Complex NWR	February 2, 2007
Sacramento RO, Sacramento River NWR Complex, SF Bay NWR Complex	February 12-16, 2007
Portland RO, Oregon Islands, and Willamette Complex	February 12–16, 2007
Hadley RO, Parker River, Silvio Conte,	February 12-16, 2007
Cabeza Prieta, Buenos Aires, RO -Albuquerque	Feb. 26 – Mar. 2, 2007
Blackwater NWR (Chesapeake Marshlands Complex)	March 12, 2007
Anchorage RO, Kenai, Alaska Maritime NWRs	March 20-30, 2007

- *Partners and State Fish and Game Surveys*: [4] Two additional surveys were conducted to solicit the views of Refuge System partners on the quality of their partnerships with the Refuge System and on their views of the Refuge System's effectiveness. These surveys were:
    - A survey of local Partners and Friends Groups, which was undertaken from March 17-25, 2008. A total of 83 responses were received from 98 potential respondents. The response rate was 85%. Most respondent were Friends Groups, but the survey also included several Audubon Society operations that serve as Friends Groups.
    - A survey of officials from state fish and game agencies was conducted from April 29–May 16, 2008. Responses were limited to one response per state agency. A total of 32 states responded to the

survey, and the Alaska Department of Fish and Game sent a letter on their views of the relationship with the NWRS in lieu of completing the survey. The thirty-two responses received constitute a response rate of 64%. However, many of the respondents did not complete all of the survey's questions and the response rate for most questions was relatively low – often there were 18 or 19 responses for each survey question. There were very few responses provided to open-ended questions.

MSI's evaluation team included:

- David Callihan, Team Leader and Senior Evaluator
- Keith Brown, Senior Evaluator
- Neal Sigmon, Senior Evaluator
- Whitney Tilt, Senior Evaluator
- Barbi Broadus, Contract Manager and Research Assistant
- Dennis Marotta, Research Assistant
- Amanda Stark, Research Assistant.

The next chapter presents the Refuge System's operating context and is followed by chapters analyzing each of the Refuge System's twelve strategic outcome goals.

THE REFUGE SYSTEM'S MISSION

The mission of the National Wildlife Refuge System is to administer a national network of lands and waters for the conservation, management and where appropriate, restoration of the fish, wildlife and plant resources and their habitats within the United States for the benefit of present and future generations of Americans. To accomplish its mission the Refuge System has developed a strategic plan that contains twelve discrete strategic outcome goals (SOGs). These objectives cover the areas of habitat and wildlife conservation, wildlife-dependent recreation, law enforcement, fire management, conservation planning and organizational excellence, among others. This evaluation report reviews each of the Refuge System's twelve strategic outcome goals and provides an assessment as to how well the system is doing in accomplishing each goal. Each section analyzes the Refuge System's progress and concludes with recommendations for how the system can strengthen its future performance.

This evaluation was undertaken to provide an objective analysis of performance; however, the evaluation team would like to note that one of the strongest findings we take away from the study – and one that is not necessarily objectively verifiable or quantifiable – is that the Refuge System has a workforce that is tremendously talented and has an extraordinary commitment to its mission and to protecting the nation's wildlife. It was impossible for the evaluation team not to be impressed with the passion and commitment of the staff at all levels of the organization — together with its land base, the workforce is truly the Refuge System's greatest asset.

The National Wildlife Refuge

The US National Wildlife Refuge System (NWRS) was created by Executive Order on March 14, 1903 when President Theodore Roosevelt established the country's first wildlife refuge on Florida's central Atlantic coast – the Pelican Island National Wildlife Refuge (NWR). Pelican Island NWR was established to conserve shorebird populations, particularly egrets, that were being decimated by commercial hunters who were filling a fashion demand for bird plumes for women's hats.

From its modest beginning on Pelican Island the Refuge System has expanded into a network of over 550 distinct units that encompasses over 95 million acres. The largest refuges, the Arctic NWR and the Yukon Delta NWR, both in Alaska, are each over 19 million acres (larger than Maryland). The *National Wildlife Refuge System*, which is the world's largest system for managed and protected wildlife, is the only federal land management system created principally for the benefit of wildlife. In addition to being grand in magnitude, the Refuge System contains some of the country's most spectacular wildlife and includes habitat critical to maintaining wildlife populations, especially endangered species, migratory birds and large mammals. In analyzing the Refuge System's performance, as per its strategic plan, it is useful to keep in mind that it is really only since the passage of the 1997 Refuge Improvement Act (RIA) that there has been an organic policy in place to provide a foundation to enable refuges to be managed under a unified system. The Refuge System's transition from a collection of individual refuges to a coherently managed system is a work in progress. In many ways, the progress has been remarkable –hundreds of management plans have been developed; public consultation processes have been instituted and undertaken; and a wide range of policies have been drafted to guide decision- making and increase consistency. There is, however, still work that remains to be done to increase the system's overall effectiveness, as should be expected at this point in time.

Strategic Conservation: The routes followed by North American migratory birds generally involve north- south travel, as many birds breed and summer in the northern climates, including Alaska, and then in the winter migrate south in search of food and milder weather. The term "flyway" has come to define the major north-south migration routes in North America, of which there are four principal flyways: the Atlantic, Mississippi, Central and Pacific Flyways. Hundreds of National Wildlife Refuges have been strategically established north to south along these flyways – creating stepping stones and feeding stations to enable millions of birds to complete their annual migrations. Today, there are more refuges established for supporting migratory bird habitat than for any other purpose. (National Wildlife Refuge System Improvement Act of 1997; Public Law 105-57.

In reviewing the performance of the Refuge System's individual strategic outcome goals it is important to remember that the Refuge System is in the midst of a significant organizational cultural change and that such changes do not happen overnight. The Refuge System goals are ambitious and progress, not surprisingly, has been somewhat uneven. Uneven progress, however, should not be allowed to detract from the remarkable transition that the Refuge System is moving forward with, nor should it detract from the incredible network of refuge lands that are being protected for the benefit of the nation's wildlife.

—*Jamie Richie/FWS*- Hundreds of thousands of snow geese stop at Bombay Hook NWR during their annual migration. The Bombay Hook NWR is one of many refuges established to strategically support migratory bird conservation.

Chapter II

EFFECTIVENESS RATINGS

The following table (2) presents the overall effectiveness ratings for each Strategic Outcome Goal (SOG) contained in the Refuge System's 2007 strategic plan. An overview of the performance of each SOG, including conclusions and recommendations, is presented in later sections of this summary chapter. Detailed performance findings can be found in the main chapter

Table 2.

Ratings	Goals
Operating Context: Budgets are down over the last several, as purchasing power has declined 11% between fiscal year 2003, and 2007. During this period of budget decline the Refuge System's operational costs have increased due to inflation and annual salary adjustments. In addition, over the past five years, administrative non-core requirements have increased.	
Highly	SOG 6: Facilitate Partnerships and Cooperative Projects to Engage Other Conservation Agencies, Volunteers, Friends, and Partners in the NWRS Mission.
Effective	SOG 5: Provide Quality Wildlife Dependent Recreation and Education Opportunities.

Table 2. (Continued)

Ratings	Goals
Effective	SOG 8: Provide Infrastructure and Equipment Adequate to Support Mission and Maintained in Good Condition
	SOG 9: Complete Quality and Useful Comprehensive Conservation Plans on Schedule and with Full Engagement of Partners.
	SOG 11: Reduce Wildfire Risks and Improve Habitats.
Partially Effective	SOG1: Conserve Manage, and Where Appropriate, Restore Fish, Wildlife and Plant Resources and Their Habitats
	SOG 3: Ensure that Unique Values of Wilderness, other Special Designation Areas, and Cultural Resources are protected.
	SOG 4: Welcome and Orient Visitors.
	SOG 12: Promote and Enhance Organizational Excellence.
Ineffective	SOG 7: Protect Resources and Visitors through Law Enforcement.
	SOG 10: Strategically Grow the System
Unable to Evaluate	SOG 2: Provide Quality Environments with Adequate Water.

Chapter III

Operational Context: Budget and Workload Trends

Overall Operating Context: The confluence of declining budgets, declining staff, and a notable increase in administrative workload has impaired the Refuge System's ability to focus on and accomplish its core mission – that of conserving habitat and resources. The performance of individual strategic outcome goals, presented in subsequent sections of this summary chapter, should be viewed in consideration of the Refuge System's overall operating context.

Budget Trends: NWRS budgets have declined over the past several years, with inflation adjusted purchasing power having declined about 11% between the FY 2003 peak and the requested FY 2008 budget. Maintenance funding, however, has significantly increased – with a jump of 436% over eight years (FY 1996 – FY 2004).

Recent budget declines appear to have severely affected refuge operations. This is evident based on a number of findings, including:

- 94% of refuge managers' survey comments indicated an inability on the part of the NWRS to accomplish its mission due to inadequate budgets and staffing;
- Workforce planning exercises are leading to significant cutbacks in personnel and services; for example, the Region 4 plan calls for a 20% reduction in staff;
- In several regions, key services such as visitor programs, environmental education, and biological monitoring are being curtailed or eliminated; and
- A number of refuges are being de-staffed; for example, in Region 5, 7 of 71 refuges will be de-staffed within the next year.

(By Fiscal year in Hundreds of Millions)

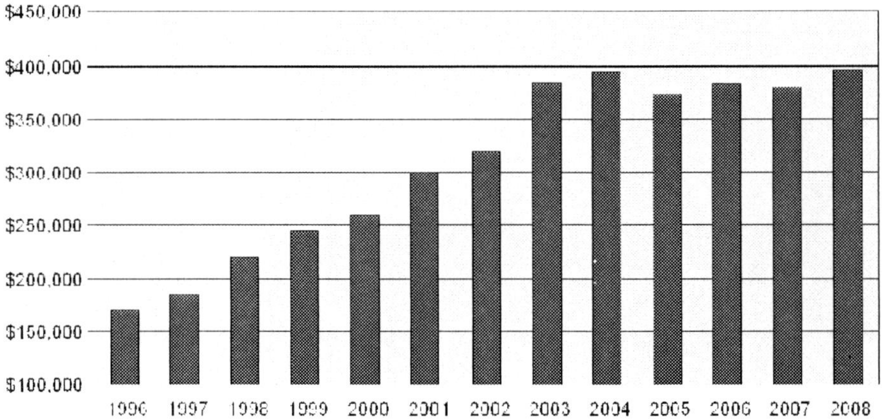

While the NWRS' actual budget has stagnated in recent years (as per above chart), inflation adjusted purchasing power declined about 11% between the FY 2003 and the requested FY 2008 budget.

Figure 1. NWRS Appropriations: Actual Dollars.

Administration/Workload: Refuge System administrative reporting has reached an unbalanced and critical level and is diverting time and resources away from mission-critical activities. There has been a clear trend, particularly over the past five years, of increased workload requirements and increased administrative reporting. While some of the workload requirements, such as the need to produce CCPs, directly support the core mission of the Refuge System, much of the work relates to administrative requirements, such as the implementation of multiple and apparently redundant timekeeping and accountability processes. Much of the effort to address accountability concerns is disproportionate to the resources involved; for example, small refuges must use the same complex systems as large refuges even though their annual operations budgets may be as a small as $20,000-$30,000 per year. The Refuge System places an emphasis on accountability that often times appears to be disproportionate to the level of resources being monitored, which is not cost effective and is a distraction from maintaining a focus on the achievement of the organization's core conservation mission.

Opportunities should be identified to reduce or streamline administrative processes. For example:

- Consideration could be given to allowing refuges with small staffs to "opt out" of certain reporting requirements;

- Reduce redundancies in data input between accounting, timekeeping and maintenance information systems;
- RAPP performance reporting could be streamlined, or modified to be more useful to management decision making;
- Onerous on-line training requirements could be re-visited, relaxed or eliminated;
- A better centralized information system could be established to help field routine requests for program information so as to reduce the need for data calls to field offices;
- Review the need to create specialist positions to handle complex administrative functions, for example the facility and asset management requirements, so that the reporting burden does not fall unduly on refuge managers (which may particularly be the case for small refuges);
- The Refuge System could recommend that specific externally required processes, such as the Federal Register listing process, be reviewed through a business process re- engineering exercise as a way to reduce administrative complexity and increase efficiency.

Accountability Concerns: Refuges with limited budgets and few staff have to live within the same rules and regulations as Federal Departments with billions of dollars at risk, such as the Department of Defense. At the Potomac River Complex, in a typical year, after salaries, about $80,000 is available to operate three refuges. Fixed costs for utilities and buildings use $60,000, which leaves only $20,000 available to support project activities. This leaves little money for waste, fraud, and abuse, but the financial controls, the checks and balances, and the multiple NWRS reporting requirements are the same as for refuges that operate with substantially larger budgets. Since the complex does not make more than 25 purchases of over $3,000 per year, it does not have warrant authority and has to obtain multiple bids (for purchases of goods and services), send proposed expenditures to the regional office, and await approval. Meanwhile, on the ground, needs are delayed and staff resources are consumed complying with rules and regulations that are appropriate when millions of dollars are at stake but seem overdone in light of the amount of funds involved.

Through the combination of a mix of some of the above mentioned opportunities, together with other actions, it may be possible to meaningfully reduce the administrative workload of key refuge management staff. Subsequent sections of this report address the NWRS' performance against the twelve strategic objective goals outlined in the 2007 NWRS Strategic Plan.

Chapter IV

ANALYSIS OF STRATEGIC OUTCOME GOALS

SOG 1: CONSERVE, MANAGE, AND WHERE APPROPRIATE, RESTORE FISH, WILDLIFE, AND PLANT RESOURCES AND THEIR HABITATS TO FULFILL REFUGE PURPOSES, TRUST RESOURCE RESPONSIBILITIES, AND BIOLOGICAL DIVERSITY/INTEGRITY

Performance Rating: Partially Effective

This objective is rated "Partially Effective" because of the significant amount of refuge land that is need of additional management attention and the inconsistent application of science-based management across the Refuge System. As per the Refuge System's RAPP performance reporting system, 89% of refuge lands – 76.5 million acres – are in Class I condition, which means the land is receiving needed management action or does not require additional management action at this time. For NWRS habitats outside of Alaska, 59% of the 18.9 million acres were reported as Class 1 in 2006. Assuming RAPP data provides an overall indication of overall habitat conditions on NWRs outside of Alaska, 41% of refuge habitats, or 8 million acres are in need of management and/or restoration to achieve the Refuge System's habitat objectives. Of refuge managers surveyed, 77% indicated that they believe their refuge is not meeting, or is only partially meeting, its habitat management goals. A majority of refuge managers (65%) also indicated they thought that staffing and budgets are insufficient to achieve the priority goals in their Comprehensive Conservation Plans.

A significant portion of refuges have not developed Habitat Management Plans, and there is an insufficient level of biological inventory and monitoring work being done –only 11% of refuge managers surveyed described the current level of inventory and monitoring work as being mostly or fully sufficient. At refuges with proper staffing and adequate budgets, this program is "effective," but many refuge units do not have adequate staffing and budgets. In addition, greater attention needs to be given to developing a better system-wide approach for monitoring habitat improvement and for better connecting the Refuge System's work with the larger mosaic of conservation lands. Despite the lack of consistent system-wide practices, the Refuge System contains many examples of outstanding habitat planning and monitoring systems. The Refuge System's challenge is to better define high-priority system-wide needs, identify best practices that meet those needs, and replicate the improved monitoring practices in an increased number of locations. Progress against this objective will be difficult to achieve with the current level of funding and staff.

CONCLUSIONS

The principal components of the Refuge System's adaptive management process, as per the Biological Management Process defined in the Refuge System's strategy, are: 1) develop habitat goals and objectives (which is part of the CCP process); 2) habitat management planning 3) management to achieve habitat objectives; 4) monitoring habitat and populations; 5) maintaining an inventory and monitoring database (including establishing baseline data). Conclusions for particular aspects of the Refuge System's adaptive management process follow.

Status of Habitat: As per the Refuge System's RAPP performance reporting system, 89% of refuge lands – 76.5 million acres – are in Class I condition, which means the land is receiving needed management action or does not require additional management action at this time. Alaska's sixteen refuges report that 98% of their habitats were in Class 1 condition in 2006, with 74,965,379 acres classified as needing no active management and 192,581 acres receiving needed management (as per 2006 RAPP data). Examining the net change of habitat conditions, as reported by RAPP, 8.4 million acres were added to Class 1A habitats from 2005 to 2006, suggesting a change in habitat classification and/or reporting capability as opposed to a change in actual habitat condition (an example of the need to exercise caution when using RAPP data).

For NWRS habitats outside of Alaska, 59% of the 18.9 million acres were reported as Class 1 in 2006. RAPP reported the addition of 1.68 million acres from Class 2 or 3 to Class 1 between 2005 and 2006. As in Alaska, this improvement likely reflects a change in habitat classification and/or reporting capability as well as a smaller change in improved habitat conditions.

Assuming RAPP data provides an overall indication of overall habitat conditions on NWRs outside of Alaska, 41% of refuge habitats, or 8 million acres, are in need of management and/or restoration to achieve the Refuge System's habitat objectives. The negative impact of these degraded lands on the NWRS' ability to conserve fish, wildlife, and plant resources is not quantified.

Habitat Management Planning: The Refuge System's emphasis on the development of Habitat Management Plans (HMPs) to guide the system's adaptive management process is not yet fully effective. This is a result of the fact that many refuges – 54%-- do not have HMPs and that current funding is inadequate to properly implement an adaptive management process across the Refuge System.

Monitoring Habitat and Populations/Inventory and Monitoring: In order to operate an effective adaptive management system of wildlife refuges the Refuge System must have the ability to conduct adequate inventory and monitoring. As a system, such capability is not in evidence. Only 11% of refuge managers surveyed indicated that current levels of monitoring and inventory work are mostly to fully sufficient to accomplish the Refuge System's mission. While budget and insufficient personnel appear to be primary reasons why inventory and monitoring work is insufficient, the absence of systematic standards and protocols is also a major contributing factor.

The Refuge System, however, does operate exemplary inventory and monitoring systems at numerous individual refuges – the HAPET system in Fergus Falls is one such example, as is the monitoring work being done at the Sacramento NWR. In both of these cases, systems are in place to identify priority species goals, identify the habitat requirements to maintain or expand species populations, and test the impact of habitat management and restoration on species populations. A challenge for the Refuge System is to learn from and document what makes these systems successful (document best management practices) and then develop a process to replicate the systems on a larger scale throughout the Refuge System. As appropriate, refuges in the same geospatial ecological regions should be encouraged to operate complementary or collaborative monitoring systems for key species.

An Example of Successful Annual Habitat Management Planning from the Sacramento NWR: Every year before the spring work cycle begins the Refuge implements its annual habitat management planning process. This process involves all key refuge staff– including managers, biologists, visitor services, law enforcement and maintenance staff– traveling together to each of the refuge's more than 40 management units. While in the field, the team discusses the issues associated with each management unit, identifies the tasks to be undertaken over the coming year, and develops a list of priorities for the year. This activity can take several weeks to complete but has proven to be an excellent process for annual planning and for ensuring that the activities of all staff are coordinated to achieve common goals. Once the field-based planning is complete a detailed written work plan is developed. The annual work plan is used for developing individual performance goals, scheduling staff time, allocating the refuge's budget and for periodically monitoring and reviewing progress.

– *David Callihan*- Refuge Manager surveys waterfowl at Eufaula National Wildlife Refuge, Alabama.

The Refuge System has invested considerable time and effort into the development of specific monitoring systems and protocols. Such systems include the Refuge Lands Geographic Information System (RLGIS), which is most active in Region 6, and the development of particular species protocols by Biological Monitoring Teams, such as the shorebird and marsh bird protocols. While all of these efforts seem promising and useful, they have do not appear to have been adopted as standard monitoring processes across the Refuge System.

In addition, momentum is gathering within the FWS in support of implementing a Strategic Habitat Conservation Initiative (SHCI), which is a planning and monitoring system that is largely based on the work of HAPET and other similar efforts. While this initiative shows promise for helping the Refuge System improve its monitoring and inventory program, the operational implementation of this program remains somewhat vague. The practices, standards, processes, and organizational support that will be required to make this initiative effective have not yet been well defined or adequately resourced.

Management to Achieve Habitat Objectives: There is not sufficient data, or analysis of the data that does exist, to draw a conclusion as to the effectiveness of the Refuge System's habitat management and restoration activity. In part, the RAPP system was designed to provide information on habitat management effectiveness but the information is not analyzed and is not presented in the context of overall system goals, targets or needs. Of refuge managers surveyed, 77% indicated that they believe their refuge is not meeting, or is only partially meeting, its habitat management goals. A majority of refuge managers (65%) also feel that staffing and budgets are insufficient to achieve the priority goals in their CCPs.

As currently constructed, RAPP imposes a top-down set of reporting definitions that are not useful for refuge-level assessment as the level of detail is too general to inform specific actions. The current RAPP system for habitat management and restoration has the potential to usefully report on system-wide accomplishments but could benefit from some modification and increased effort toward interpretation and analysis. RAPP data is currently not receiving adequate quality control review and correction.

Influence of Scale/Islands of Excellence: Many refuges are too small to achieve biological integrity, contribute meaningfully to biological diversity, or protect their own environmental health by themselves – one-half of the refuge units in the NWRS are smaller than 5,000 acres. In order for the Refuge System's many smaller units to play a critical role in species conservation, they need to become fulcrums for influencing conservation actions in larger landscapes, either through becoming catalysts for habitat prioritization and land use, or through developing innovative habitat management practices and ensuring those practices are replicated in larger landscapes. It is not clear that the Refuge System currently places adequate emphasis on this role.

RECOMMENDATIONS

Hire Additional Biologists: As noted in the conclusion section, in part, the Refuge System is unable to fulfill its commitment to manage refuges using an adaptive management system because of a shortage of biologists (approximately 20% of the Refuge System's workforce are biologists). It is recommended that the Refuge System review the adequacy of its biology workforce as compared to system needs. For example, an assessment could be undertaken on the number of biologists that are required to implement currently approved CCPs versus the number of biologists now employed by those refuges. Such a study would provide a sense of the extent of the gap that currently exists. The White Paper produced for the Conservation in Action Summit recommended that biological teams be added to the top 50 refuges. An assessment should be undertaken to determine the degree that this has happened. Consideration should also be given to adding a performance measure to RAPP reporting data under SOG 1 on the adequacy of the biological workforce.

Habitat Management Planning: A clarification should be made regarding whether detailed habitat management planning is incorporated into the CCP documents or becomes a separate stand-alone document – it is currently difficult to know the level of adequacy of habitat management planning as CCPs sometimes include habitat plans and sometimes don't.

The development and implementation of habitat management plans are one of several significant operational areas that appear to be underresourced. Given the Refuge System's plethora of planning and management requirements, and a shortage of funding to do all that is required, it is probably worthwhile to:

- Prioritize the planning and management actions that refuge managers are expected to annually undertake;
- Differentiate management requirements as per refuge and staff sizes, so that smaller refuges can have more time and resources to focus on core activities (and be excused from select other functions) and enable larger refuges to take on an expanded role and focus on an increased number of biological and monitoring priorities; and
- Better define regional biological priorities/objectives so that refuges can structure their activities to support these goals.

Inventory and Monitoring: As small budgets and limited staffs are unlikely to improve significantly in the near future, refuges need to carefully identify priorities within their capacity to conduct complete robust adaptive management pursuant to the integrity-diversity-health policy. The Refuge System should give consideration to developing a higher level of monitoring and inventory standardization among its refuges and, in part, focus those systems toward the management needs of regional priorities. At the current time, there is a substantial amount of monitoring and inventory work taking place but the effort has not been well coordinated or standardized. In addition, effort toward developing a system-wide geographic monitoring capability should be continued and provided increased emphasis, e.g. explore the feasibility of using RLGIS more widely throughout the system.

Scale of Influence: As refuge appropriations represent only a small portion of the funding available for conservation, NWRs must continually engage and broaden partnerships with states, tribes, federal agencies, and others to combine existing funding for priority projects and create new funding opportunities (see Partnerships section). Incentives, measurement systems, and planning processes should be developed to ensure that increased emphasis is given to the Refuge System's role in influencing larger conservation landscapes through planning leadership and developing habitat management approaches that can be replicated by others. This beyond the boundaries perspective is also required to conduct large scale planning and identify regional conservation priorities at the landscape level (as discussed under SOG 10 - Strategic Growth of the NWRS). Management actions and performance measures should be adopted to ensure this issue receives adequate emphasis and analysis.

SOG 2: Provide Quality Environments with Adequate Water –Refuges/WMDs Have Clean Air, Water, and Soils (Meet Federal and State Standards) and They Have Ready Access to Adequate Quantities of Water to Fulfill the Purposes of Each Refuge and the Mission of the NWRS

Performance Rating: Unable to Evaluate

This objective is rated "Unable to Evaluate" as a result of the limited information available against which to undertake an assessment of this strategic goal. It is recommended that the Refuge System work to develop a water management strategy and policy in order to provide increased support to water issues, including addressing issues of water rights, quality and quantity.

CONCLUSIONS

In general, the MSI evaluation team found it difficult to assess air and water conditions and the level of impact from contaminants at the NWRS level. The data collected by RAPP has limited value for assessment purposes due to the nature of data collected, the lack of consistency of data collected, and overall usefulness of the information collected. Three examples, displayed in Table 3, are illustrative of the difficulties encountered.

Table 3.

Issue	2005 RAPP	2006 RAPP
Changed Indicator	(2.03) 303d-listed waters = 197	(2.04) On-refuge acres of State 303d-listed water = 488,398
Suspect Data	(2.05) Water resource assessments conducted = 149 (2006 target = 204)	(2.06) Water resource assessments conducted = 399
Usefulness of Information Collected	303d-listed waters as a NWR metric is of limited value since NWR has little control over designation and remediation, and the impact of these waters on the Refuge Mission is not quantified.	

Water: Many refuges face critical water flow concerns, as well as water quality issues. As regional demands for water rise, select refuges will face increased demand for water they already rely on, and increased difficulty in securing new sources. The NWRS is directed to be an advocate for its trust resources in the adjudication and allocation of water rights, but they face heavy competition for these water resources from a wide range of agricultural and development interests nature of data collected, the lack of consistency of data collected, and overall usefulness of the information collected. Three examples, displayed in Table 3, are illustrative of the difficulties encountered.

Overall, the Refuge System does not currently operate a well defined and structured water resources program, as there is a lack of guidance as to how refuges should be addressing water rights issues and what information should be collected. In addition, there is no individual or office designated to coordinate the program.

Air Quality: Along with other areas managed by the National Park Service and U.S. Forest Service, the Refuge System's 21 Class 1 airsheds provide an important national barometer for overall air quality and the nation's growing interest in air visibility. Continued strong cooperation with the National Park Service, U.S. Forest Service, and states will be critical for viable monitoring and development of appropriate solutions, especially in the current fiscal climate.

Contaminants: The evaluation team was unable to determine the adequacy of the Refuge System's management of contaminants.

RECOMMENDATIONS

Develop a Water Strategy: The Refuge System should develop an overall strategy and management structure to more effectively assess and address water management issues. Steps to develop such a program would include:

- Convene a working group to better define the management of a water resources program.
- Appoint a Water Resources Coordinator, who would work full-time on Refuge System water rights issues.
- Develop a policy, or at least a defined process, for how refuges should assess water rights needs; define what information should be collected and how it should be catalogued; construct meaningful RAPP reporting measures, which will provide an indication of system progress; and develop an inventory of unresolved water issues and required solutions, by refuge unit.

In addition, the Refuge System should ensure that these issues are addressed in existing or pending CCPs and progress should be reviewed and reported on an annual basis.

Develop an Operational Water Management Program: Once an overall water management strategy is developed, the Refuge System should create a staffing structure and budget to support required implementation actions.

SOG 3: Ensure That Unique Values of Wilderness, Other Special Designation Areas, and Cultural Resources Are Protected

Performance Rating: Partially Effective

The Refuge System is currently operating under a 1986 Wilderness Policy, which needs to be updated in consideration of the 1997 RIA and to provide better guidance on the actions that are appropriate for managing wilderness areas. A new draft policy was developed and released for public comment in 2001, but the policy has never been finalized and the process that has been used to finalize the policy has been inefficient and lacks transparency. At this time, there is a lack of systems for managing or tracking wilderness lands system-wide; for example, there is no central repository of Wilderness Management Plans, no documentation on threats and violations to wilderness areas, and no information available on minimum requirements analysis, which are required of refuge managers to determine appropriate wilderness management actions. The NWRS, however, has supported the development of wilderness training courses and refuge managers overwhelmingly feel these courses have been effective in enabling them to acquire the skills necessary to manage wilderness areas. In addition, efforts are currently underway to support the development of an interagency wilderness monitoring protocol. Despite the shortcomings in policy development and information systems, the NWRS has successfully addressed the most important factor for managing wilderness areas: it has provided the training necessary to ensure most on-the-ground managers have adequate wilderness management skills.

Conclusions

Policy Development: The current NWRS Wilderness Policy is outdated and needs to be revised. The process used by the Department of Interior to update its wilderness policy has been inefficient and has lacked transparency. The lack of an updated Wilderness Policy hinders the NWRS' ability to manage wilderness areas in a consistent manner and has prevented the establishment of clear guidelines for conducting minimum requirements analysis.

Wilderness Management Planning and Monitoring: The degree to which refuges have acceptable and consistent Wilderness Management Plans is unclear as there is no central repository of these plans and thus there is no way to review the adequacy of these plans on a system-wide basis. In addition, there is not sufficient monitoring data, or a sufficient monitoring system, to track the status and quality of wilderness conditions within the NWRS. However, effort is currently underway to develop an interagency wilderness monitoring system, and the NWRS has played a lead role in funding this effort. It is expected that monitoring protocols will be available for review sometime in the current calendar year (FY 2007). The RAPP system data does not appear to be of much value for understanding wilderness decisions or informing management decisions, nor is the data utilized other than for reporting.

Training: The NWRS has played a key role in supporting the Carhart National Wilderness Training Center and in developing and delivering wilderness training. Through this interagency approach, refuge managers are able to access high- quality wilderness training courses that provide them with the skills they need to manage wilderness areas. The training provided is considered to be effective and a majority of refuge managers responsible for managing wilderness areas have completed wilderness training (64% of refuge managers who manage wilderness areas indicated they have completed the required National Wilderness Stewardship Course).

RECOMMENDATIONS

Policy Development: Given the extended time period since the draft wilderness policy was released for public comment (about six years), it is recommended that the NWRS re-release the proposed wilderness policy for an additional round of public comment prior to its finalization.

Wilderness Management Planning and Monitoring: The NWRS should increase its ability to collect information on wilderness conditions, threats, and best practices, particularly regarding minimum requirements analysis. Once the new wilderness monitoring protocols are developed by the interagency coordination group these indicators, or at least a sub-set of proposed indicators, should be considered as reporting measures for the RAPP system.

Training: The NWRS should continue to support the interagency training process being implemented through the Carhart National Wilderness Training Center. Additional effort should be made to ensure that a higher percentage of refuge managers who are responsible for managing wilderness areas are able to complete the National Wilderness Stewardship Course. The number of managers completing this course should be considered as a performance measure in this area (as part of a revised RAPP system).

SOG 4: Welcome and Orient Visitors

Performance Rating: Partially Effective

The NWRS operates its public use program based on a set of visitor use standards, including standards that focus on the orientation and welcoming of refuge visitors. Indications are that NWRS performance against these standards - which include, for example, guidelines for appearance and placement of signage, brochure and publication formats, and website design protocols - has been improving (RAPP data show that the number of field stations meeting each of the seven related standards increased between 12% and 29% per standard from 2005 to 2006). Despite these recent improvements, a substantial portion of refuges and wetland management districts are not currently meeting standards related to welcoming and orienting visitors. For example, depending on the data source, between 33% and 47% of field stations have inadequate or inappropriate signage. Similarly, approximately one-third of refuges have websites that do not meet NWRS standards, are not current, or are deemed by the relevant refuge manager to be insufficient. Somewhat in contrast to this general NWRS picture is the situation at high visitation refuges, and more specifically, high visit refuges with visitor centers and comparatively well-developed visitor programs. For this group of NWRS field stations, informative brochures and publications are readily available; signs are useful, adequate in number, and appropriate in placement; and visitor interactions with staff and volunteers are overwhelmingly positive – i.e., courteous and informative. On all of these factors, and a number of others, greater than 90% of surveyed visitors indicate strong performance by this specific category of refuges. In summary, the NWRS is moving in the right direction with regard to welcoming and orienting visitors, but it still has a ways to go.

Conclusions

The Refuge System's performance with regard to facilitating visitor access to refuges is uneven: Approximately one-third of NWRS websites include neither directions nor maps that would help visitors find specific refuges. Similarly, one-third of refuge managers feel signage is insufficient in terms of helping visitors find their refuge or wetland management district [5]. High visitation refuges may represent an exception to this situation – visitors to these refuges indicate signs and other materials have done a good job of helping them find the refuge they are visiting.

Related to the first conclusion, the current standards for signage and websites do not fully address this objective of the Refuge System: Sign standards provide little, if any, guidance regarding how off-refuge signs should be used and/or placed to facilitate visitor access to a refuge. Similarly, both the current and soon-to-be released FWS web standards include no requirement that refuges incorporate maps, directions, and other information that would assist individuals trying to visit a refuge or WMD (only a "contact us" function is required).

NWRS is reasonably effective in terms of informing and engaging refuge visitors but could easily improve its performance in this area: Brochures are generally informative and available at refuges, and refuge employees and volunteers are able to provide helpful and informative answers to visitor questions. However, videos and CDs – very engaging and effective means of providing information to refuge visitors - are substantially underutilized. The information provided on refuge websites is very inconsistent from refuge to refuge and frequently provides only the most basic information.

The NWRS is not, in many instances, consistent in its public messages or appearance: Refuge websites present perhaps the best illustration of this conclusion. Refuge websites vary widely in appearance, format, and content, and it is highly unlikely that a visitor to multiple NWRS websites would have the sense that they have just viewed the websites of entities within a single organization. By contrast, brochures and related published materials (e.g. maps) do present a common appearance and a single public identity.

As a corollary to the preceding conclusion, refuge websites are currently inconsistent in appearance, provide widely different types and levels of content, are not always updated, and are underutilized as a means of providing information and engaging the public. Many NWRS shortcomings related to SOG 4 could be at least partially addressed by improved and more consistent NWRS websites.

Refuge staffs are fully meeting the objective of interacting with visitors in a professional, courteous, and helpful manner.

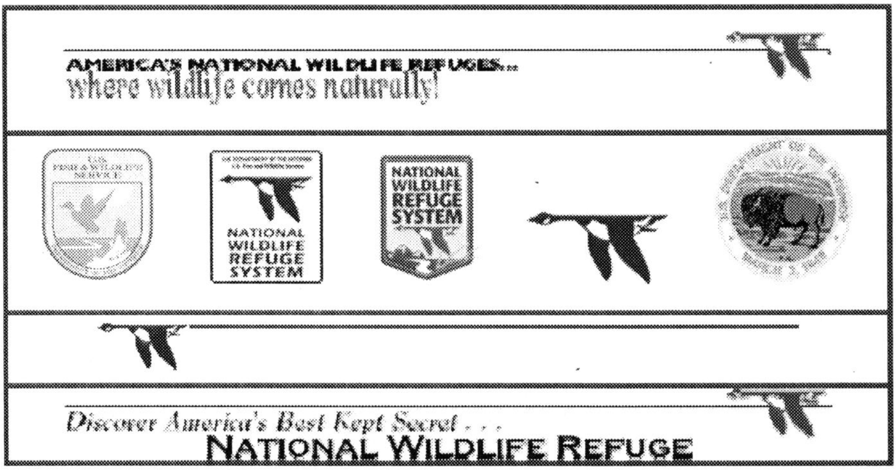

Figure 2. Selection of Logos and Slogans from Refuge Websites.

Recommendations

Website Formats: Develop a single website format/architecture for each refuge and WMD website. A working team consisting of field, regional office, and Washington office staff should develop a website "template" that can be used by each NWRS field station to create refuge-specific websites. Though not intended to be an exhaustive list, any website template should at least include the following [6]:

- Use of the same logos and slogans (including placement);
- Presentation of maps and directions to help visitors find the refuge (i.e., not just maps of the refuge lands alone) and days/hours of operation;
- Description of the main elements of the public use program;
- Description of the refuge purpose and objectives; and
- Description of the national-level mission and goals of the NWRS.

There are several options available in terms of the approach used to manage specific refuge websites; however, the most efficient option would be to centralize the management function in a single office or under a single contract. Placing responsibility for website development and maintenance in a single office will help to ensure consistent format and appearance, will greatly facilitate development and rollout of new website features and options system-wide, and will increase the efficiency and effectiveness of on-going website maintenance. While a central website office would have responsibility for placing updated content on specific refuge websites (new content would be provided by each refuge), the Refuge System website "template" could include a page that would allow individual refuges to post refuge- specific information, e.g. upcoming public events, etc.

Video Production: Produce a high quality video presentation that describes the national mission and goals of the NWRS, as well as the principal programs and characteristics of the Refuge System. Distribute the video to all field stations and encourage its presentation at visitor centers, during school visits (at the refuge or in the schools themselves), at community meetings, etc. A summary clip from the complete video presentation should also be made available so that refuges can incorporate a brief piece on the national system in any refuge-specific videos they produce.

[Note: Subsequent to making this recommendation the assessment team was told that there is a national 11- minute video of the Refuge System that has been produced for use throughout the system. However, the team did not view this video and it was not available at any of the refuges visited during site visits, nor was it ever mentioned by field staff. Assuming the video does exist, and is current and of high quality, then it should be distributed and promoted for use at refuges, and particularly for use at refuges that have public facilities to show videos and have not produced their own video.]

Review and revise standards and guidance related to visitor services to more clearly support the achievement of SOG 4: For example, provide improved guidance related to the placement of off- refuge signs and, as per the first conclusion under this section, develop a website template with specific guidelines related to format and content.

Signage: Given that refuge field staffs have clearly identified both on and off-refuge signage as a current need, the number of refuge (directional) signs should be increased.

SOG 5: Provide Quality Wildlife-Dependent Recreation and Education Opportunities

Performance Rating: Effective

The Refuge System has done a good job at expanding the number of refuges that offer wildlife-dependent recreation opportunities and, overall, the visitor satisfaction rate at refuges is very high –above 90% in the 2002 and 2004 surveys. Among the six mandated wildlife dependent recreation activities, hunting, fishing, wildlife viewing, and photography programs are widely available and adequately run. The Refuge System needs to make a concerted effort to improve its environmental education and interpretive programs, as these programs are not well defined or supported and are not able to adequately meet public demand (as per refuge manager comments). The Refuge System has taken steps to begin to develop an environmental education strategy, but additional clarity, guidance and resources are needed to increase the program's effectiveness.

Conclusions

The NWRS received nearly 35 million visitors in 2006. For the most recent years in which visitor satisfaction surveys were conducted (2002 and 2004), visitor satisfaction levels were very high, with general visitor satisfaction levels being above 90% in both surveys. Overall, the Refuge System has done a good job of increasing recreational opportunities – over 90% of the refuges that are able to provide wildlife-dependent recreational opportunities now do so. Also, the number of refuges offering hunting, fishing, and environmental education programs has expanded steadily over the past 5-10 years.

In terms of the individual Big 6 recreational activities, the operation of hunting, fishing, wildlife viewing, and photography programs are generally operating at a satisfactory level in terms of the Refuge System's ability to provide an adequate level of service and in terms of the support provided to those programs by the Refuge System. The environmental education and interpretive programs, on the other hand, are not able to meet public demand and are not adequately supported by the Refuge System. This latter conclusion is based solely on the view of refuge managers: 55% of refuge managers surveyed indicated they are not able to adequately meet the demand for environmental education services and 48% indicated they are not able to meet the demand for interpretive services.

Conclusions on particular aspects of the wildlife- dependent recreation program follow.

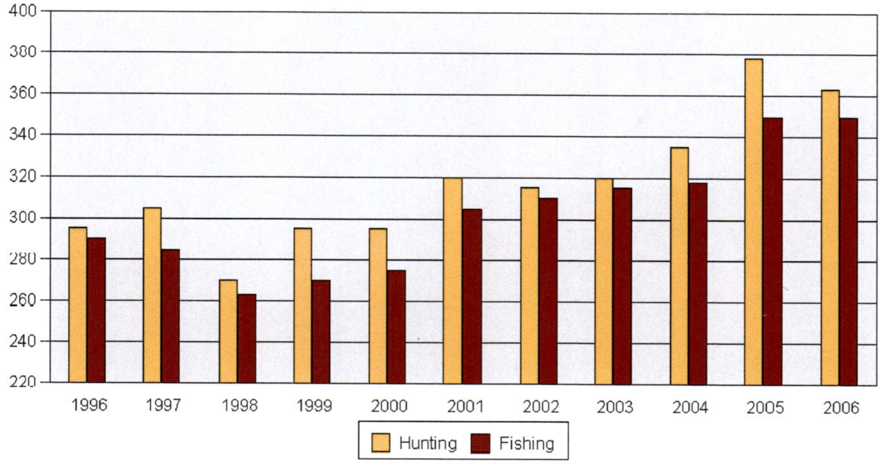

Figure 3. No. of Refuges Offering Hunting and Fishing.

Visitor Services Planning: The NWRS is not doing an adequate job in meeting its policy requirement that all refuges develop Visitors Services Plans (a requirement that was elevated in importance by the 1997 Refuge Improvement Act). Currently, two- thirds of refuges do not have such plans; of the one-third of refuges that have VSPs a majority of these plans (about 70%) were produced prior to 2002. In addition, there has been no system-wide review of the adequacy of the plans that do exist. While most refuges do not have formal Visitor Services Plans, most do have hunting and fishing plans, which are key components of a Visitor Services Plan.

Perhaps more significantly, there is no system-wide Visitor Service Plan or strategy for the Refuge System. In a time when many refuges have recently faced staffing and budgetary reductions it would be useful to be able to provide regions and individual refuges guidance as to system-wide priorities and objectives for the Visitor Service Program.

Hunting and Fishing: The Refuge System has done a good job at increasing hunting and fishing opportunities over the past decade. Between 1996 and 2006, the number of refuges that offer hunting increased 24%, to 366 refuges, and the number that offer fishing increased 22%, to 352 refuges. Also, 85% of refuge managers indicate that they are generally able to fully able to meet public demand for hunting, while 84% express similar sentiments regarding demand for fishing.

Per RAPP data, 77% of refuges indicate that their hunting programs meet a high level of quality as determined by a set of twelve quality criteria standards; the fishing program scores somewhat lower at 59%.

Partners gave the Refuge System high marks for their effectiveness in providing hunting and fishing opportunities.

- 64% of partners rated the Refuge System as being very effective to highly effective in providing hunting opportunities; 95% of partners provided a rating of between moderately to extremely effective.
- 64% of partners rated the Refuge System as being very effective to highly effective in providing fishing opportunities; 94% of partners provided a rating of between moderately to extremely effective.

State fish and game agencies provided generally favorable views regarding the Refuge System's effectiveness in providing hunting and fishing opportunities.

- 34% of partners rated the Refuge System as very effective or extremely effective at providing recreational hunting opportunities; 50% survey respondents rated effectiveness for this area as moderately effective.
- 43% of partners rated the Refuge System as very effective or extremely effective at providing recreational fishing opportunities.
- 29% of respondents rated the system's ability to provide fishing opportunities as "somewhat ineffective."
- The views of state agencies are significantly below those of partners in regard to the Refuge System's ability to provide quality fishing opportunities: 94% of partners rated this capability as between moderately to extremely effective, whereas state agencies provide a rating of 71%.

A number of comments were received from partners on the effectiveness of various programs; however, as reported in the partnership section, many of the comments dealt with the inadequacy of resources and funding and the affect this has had on the Refuge System's effectiveness. All refuges that provide hunting and fishing opportunities have hunting and fishing plans, which are a significant part of an overall Visitor Services Plan. These plans are required to adhere to National Environment Policy Act (NEPA) requirements and, as such, have been developed in consideration of public input and stakeholder views.

—*Hester, Eugene*- Between 1996 and 2006, the number of refuges offering hunting increased 24%, to 366 refuges.

Environmental Education: On a national level, the Refuge System operates a highly decentralized and somewhat ad hoc environmental education program – which is often legitimately characterized as a "demand-drive" system. The Refuge System does not have adequate policies, staffing, or strategy in place to operate an effective nation level environmental education program.

While there are numerous examples of highly impressive and innovative programs at individual refuges, there is no process for analyzing and disseminating best practices, nor is there any way to judge how well the program is operating overall, as there is no defined target audience, objectives, or standards. In addition, a majority of refuge managers indicate they are not able to adequately meet public demand for environmental education services, and a majority indicated they do not receive adequate progra m support to operate environmental education programs. Among refuge managers, the environmental education program was considered to be the least well supported program of any of the six mandated wildlife dependent recreation programs, and this was particularly true fo r smaller refuges.

A further consideration regarding environme nta l education is that programs can be relatively expensive to develop and operate. Thus, while the Refuge System has demonstrated it is able to offer exceptional environmental education program s it is not clear what the proper balance or tradeoff should be in terms of dedicating tight resources to environmental education versus dedicating resources to other Refuge System priorities. At the current time the level of resources and attention provided to environmental education appears to be somewhat arbitrary. In or around 2001, the Refuge System developed a White Paper on environme nta l education. The paper defines different tiers of environmental programs as being appropriate forenvironmental programs as being appropriate for different refuges, but does not present crite ria as to which refuges should offer which tier of program, or which refuges should not offer environmental education programs.

Partners gave the Refuge System generally marks for their effectiveness in providing environmental education and interpretation services. Fifty-four percent of partners rated th Refuge System as very effective or extremely effective at providing environmental education and interpretation activities; however, 34% rated this capability as moderately effective.

—*John and Karen Hollingsworth* - Environmental Education at Lake Woodruff NWR.

Interpretive Programs: Based on the Refuge System's basic information outreach capa bility and its challenge in meeting the demand for interpretive services, the system as a who le is under-performing in providing adequate interpretive services. In large measure this conclusion is based on the views of refuge managers as 48% of managers said they are not able to adequately meet public demand for services and 41% said they do not receive adequate technical support and guidance to manage such programs. These figures are offset by the RAPP data, which indicates that 68% of the 286 refu ges (as of FY2006) that do operate interpretative programs have "high quality" programs. A significant number of refuges do provide a wide variety of interpretive services, such as kio sks, refuge video s and even guided-tour radio programs.

This objective is closely intertwined with SOG 4– Welcoming and Orienting Visitors, which incl udes providing visitors basic information about a refuge's program and purpose (including information on its species and habitat goals) The following conclusions relate to interpretive activities (and the first tw o are carried-over from the analysis of SOG 4):

- NWRS is reasonably effective in terms of informing and engaging refuge visitors but could easily improve its performance in this area. Brochures are generally informative an available at refuges, and refuge employees and volunteers are able to provide helpful and informative answers to visitor questions. However, videos and CDs – a very engaging and effective means of providing info rmation to refuge visitors – are substantially underutilized. The information provided of refuge websites is very inconsistent from refuge to refuge and frequently provides the most basic information.
- Refuge managers expressed concern that they are unable to adequately meet the public's demand for interpretive services, with 48% percent saying they are operating at a level below that of being able to "generally meet demand." Refuge managers also expressed concern that they do not receive adequate program support to operate interpretive programs, with 41% indicating that support is below adequate as compared to 19% saying that support is above adequate or fully adequate.
- Refuge websites are currently inconsistent in appearance, provide widely different types and levels of content, are not always updated, and are underutilized as a means of providing information and engaging the public. Many NWRS shortcomings related to SOG 4 could be at least partially addressed by improved and more consistent NWRS websites.

Wildlife Viewing and Photography: In gen era l, the level of wildlife viewing and photography opportunities that currently exist on refuges seems adequate and probably does not require significant additional focus. The one caveat is that about one- third of refuge managers indicated that the technical support and program guidance they receive in this area is less than generally adequate, which would indicate opportunity for improvemen t in this area. The Refuge System can probably do a better job in expanding opportunities for wi ld life viewing and providing better guidance and resources to its managers in how this can be done. This is an area that probably offers one of the most cost-effective opportunities for the Refuge Systemate its message to a significant number

Demographic Trends of Wildlife-dependent Recreation: There are a couple of changing trends that should be considered by the Refuge System as it continues to develop system-wide priorities and programs: 1) non-white ethnic groups are an expanding part of the U.S population and are under-represented among Refuge System visito rs, for example 25% of the US population is either Hispanic or African-American but these groups represent a total of four percent of refuge visitors; and 2) wildlife-dependent recreational uses are changing, as hunting activities are on the decline in the U.S. and non-consumptive activities are on the increase. The latter category includes activities such as nature walks and bird watching. The R efuge System is in the process of developing a partnership-based Birding Initiative but, in ge neral, probably needs to give greater attention to maintaining its public support and relevance in relation to these changing trends.

RECOMMENDATIONS

Prioritize Services: In light of a high public demand nd the Refuge System's limite d and stretched budgets, the Refuge System should prioritiz e the services it will offer and provide some guidance to refuges and regions as to how limite resources should be allocated among the various recreation al activities. Region 5's experience in developing two priority Big 6 uses for each of its refuges could be instructive for how a prioritiza tion process could be developed for other regions.

Visitor Services Planning: A more concerted effort should be made to complete Visitor Services Plans for all refuges. However, this effort should probably be preceded by the development of a clear program strategy, guidance, and objectives as to what the Visitor Service Program hopes to accomplish and how refuges should prioritize visitor services (in light of limited resources). In particular, the strategy should:

- Set guidance on how to prioritize the types and levels of programs that can be offered at different refuges, based on refuge characteristics and available staff and budget;
- Identify priority Big 6 uses for each refuge; and
- Develop and distribute a simplified Visitor Services Plan template along with examples of a usefu l and not too complicated completed plan

Hunting and Fishing: There are no major recommendations in this area, although it wou ld be useful to conduct a targeted evaluation of the satisfaction level of the Refuge System's hunting program. The wildlife-dependent recreation policy does mention that the Refuge System should offer "quality" hunting experiences, but this has never been defined by the system. Defining the type of "quality" hunting experience the Refuge System endeavors to promote provides an opportun ity to further define and improve refuge hunting programs and to distinguish a refuge hunting experience from that available on other lands.

—*John and Karen Hollingsworth* - Refuge staff assisting young Fishermen at San Francisco Bay NWR.

Environmental Education: Significant attent ion is needed to strengthen the Refuge System's reas. Specific actions for consideration include:

- Establish a dedicated environmental education support team/unit to develop programs in this area and provide leadership to the Refuge System. This unit should be headed by an experienced professional who has environmental education credentials or, at a minimum, extensive experience and expe rtise. The unit does not necessarily need to be located in the Washington office.
- *Develop an environmental education strategy*: The Refuge System currently has a Cooperative Agreement (CA) with the University of Wisconsin-Stevens Point to develop a Refuge System environmental strategy – the CA runs through December 2008. In the interim, however, the Refuge System itself can become better organized and begin to provide greater strategic direction to the program. This would include a clar ification of the terms and programs related to interpretation and environmental educati on and developing a set of guidance or core activities for each program area.
- *Develop guidance on appropriate levels of programming based on refuge profiles and staff/resource availability*: One step in this process might be to develop case studies of successful environmental education program s that can be managed by differing staff and resource levels. For example, profiles could be put together of a half-dozen environ me nta l education programs operated at large, medium, and small refuges, and best practi ce guides could be developed and tailored to staffing and resources. This information c ou ld also help the Refuge System to prioritize where it offers environmental education programs and identify what types of progra ms it can offer– for example, due to cost and staffing issues the Refuge System may wis hto define which refuges are appropriate for offering EE programs, and which are not, as a way to guide resource allocation decisions.
- *Engage more widely with potential partners to develop research-driven EE programs*. Partne rs could include universities, NGOs, other DOI agencies, and various specialized groups, such as the North American Association for Environmental Education. This effort should result in criteria, standards and basic materials to guide the EE program.
- *Interpretive Programs:* Develop environmental interpretation support packages for refuges (particularly for smaller refuges). Using the US Park Service model of "discovery backpacks" [7] or other similar strategies,

environmental resources/toolkits should be developed for use by refuges throughout the system. In particular, simple self-guided activities and toolkits should be developed for use by refuges that do not have dedicated full-time environmental education or interpretive staff.

- Development of national interpretive resources, for example the development of "discovery backpacks", could be a cost- effective way to provide interpretive tools and support to refuges of all sizes and to ensure that refuges have the ability to host interpretive groups even if they do not have the time and resources to develop the materials themselves. This would be an example of a cost-effective self-guided interpretive activity that could be offered at a significant number of refuges. Centralizing the management of this program function would likely lead to increased efficiency in program development and enable the Refuge System to offer standard materials at a greater number of refuges.

SOG 6: Facilitate Partnerships and Cooperative Projects to Engage Other Conservation Agencies, Volunteers, Friends, and Partners in the Refuge System Mission

Performance Rating: Highly Effective

This objective was rated highly effective for several reasons: over the past ten years the Refuge System has been able to significantly expand participation by volunteers and Friends Groups; partnerships with thousands of local and national organizations make a significant contribution to the accomplishment of the Refuge System's key objectives, particularly in the areas of habitat restoration and visitor services; and partnerships bring a tremendous amount of funding into the system – in 2005 alone the total value of partnership contributions to the Refuge System exceeded $50 million, with over $30 million of the total being in direct cash contributions. Although the level of volunteer support has increased dramatically over the past ten years – from 383,983 hours in 1987 to 1,478,797 in 2005 – volunteer hours have somewhat declined in recent years as the system's capacity to manage volunteers has likely been reached (with the limits of current budgets and staff). Given the value of volunteer and partner contributions to the system, it is recommended that the Refuge System explore ways to assign dedicated staff to manage volunteer programs in locations where doing so is likely to be the most cost effective.

Conclusions

The partnership and volunteer SOG is considered highly effective, for the following reasons.

Partnerships are responsible for critical contributions to the achievement of NWRS goals: Partnerships operate across a wide range of refuge activities and make a critical contribution to NWRS accomplishments, including in land acquisition, habitat management, (especially wetlands restoration), monitoring and research, and environmental education. In 2005, the total value of partnership contributions to the Refuge System exceeded $50 million, with over $30 million of the total being cash contributions and the rest received in in-kind support. According to the NWRS' own data, 100% of habitat restoration on refuges in 2006 was the result of collaborative projects with partners.

(Number of Hours).

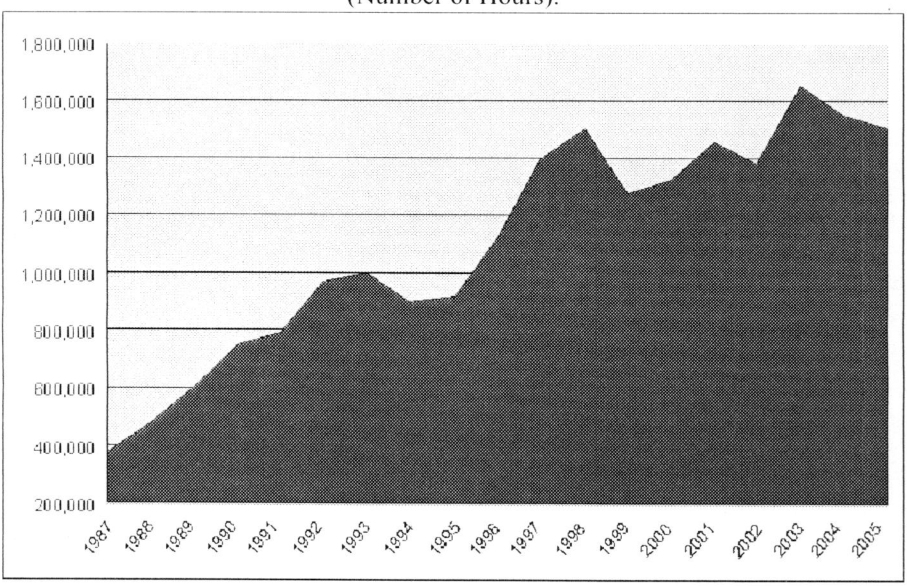

Figure 4. Annual Volunteer Hours Contributed to the Refuge System [8].

The use of volunteer support has increased significantly over the past twenty years: The Refuge System has fostered and benefited from tremendous growth in numbers of volunteers. In 1980, less than 5,000 volunteers contributed services to refuges, but by 2006 that number had grown to over 32,000. In 2005, volunteers were active on 423 stations and contributed approximately 1.5 million hours of assistance that were valued at approximately $25 million. Partnerships and volunteers supplement refuge staff, thus increasing the amount of work that gets done on refuges. In recent years, the use of volunteers has somewhat declined in concert with fluctuations and declines in the NWRS budget.

There has been impressive growth in Friends Groups and the contribution they make to the Refuge System: Since 1994, the number of Friends Groups has expanded from 75 to approximately 250. These organizations represent an estimated membership base of more than 40,000. The NWRS support of these groups through Refuge System investments in training, grant programs, workshops, and mentoring has been responsible for their expansion, which has been a great asset to the Refuge System.

National Organizations: The NWRS has close and productive partnerships with a number of major national conservation organizations, including Ducks Unlimited, the Nature Conservancy, the Audubon Society, the Conservation Fund, the Trust for Public Land and others. The NWRS also benefits from funds provided by the National Fish and Wildlife Foundation, which provides tens of millions of dollars in grants each year for projects that benefit fish and wildlife management, often in collaboration with refuges.

Table 4.

Partner Survey Data: Summarize the Overall Quality of the Working Relationship between your Organization and the NWRS					
	Poor	Fair	Good	Very Good	Excellent
Regional/Washington Office	7%	13%		27%	25%
	21%		27%	52%	
Individuals Refuges	3%	6%		24%	56%
	8%		13%	80%	
Total no. of responses = 67 for Washington/regional Office; 72 for local refuge.					

Data from Partners and Friends Groups Survey: Data from the Partners Survey indicated a high degree of satisfaction among partners with both the quality of their partnership with the Refuge System and with their overall view of the effectiveness of the Refuge System. The following tables present data from the MSI Partners and Friends Groups Survey (Partner Survey), conducted March 17-25, 2008.

As can be seen from the above table 4, a majority of partners who responded to the survey rated the quality of their relationship with the Refuge System as very good to excellent.

- 52% of partners rated the quality of their relationship with the regional/Washington office as excellent or very good; and
- 80% of partners rated the quality of their relationship with individual refuges as excellent or very good (and 56% rate the quality of the relationship as excellent).

Data from the State Fish and Game Agency Survey: Data from this survey also indicated a high degree of satisfaction among state fish and game agencies with both the quality of their partnership with the Refuge System and with their overall view of the effectiveness of the Refuge System. The following table 5 presents data from the State Fish and Game Agency Survey, which was conducted April 29-May 16, 2008.

Table 5.

<table>
<tr><td colspan="6">State Fish and Game Agency Survey Data: Summarize the Overall Quality of the Working Relationship between your State Agency and the NWRS</td></tr>
<tr><td></td><td>Poor</td><td>Fair</td><td>Good</td><td>Very Good</td><td>Excellent</td></tr>
<tr><td rowspan="2">Regional/Washington Office</td><td>7%</td><td>7%</td><td rowspan="2">43%</td><td>14%</td><td>29%</td></tr>
<tr><td colspan="2">14%</td><td colspan="2">43%</td></tr>
<tr><td rowspan="2">Individuals Refuges</td><td>0%</td><td>12%</td><td rowspan="2">41%</td><td>29%</td><td>18%</td></tr>
<tr><td colspan="2">12%</td><td colspan="2">47%</td></tr>
<tr><td colspan="6">Total no. of responses = 17 for Washington/regional Office; 17 for local refuge.</td></tr>
</table>

As can be seen from the above table, a majority of state agencies who responded to the survey rated the quality of their relationship with the Refuge System as between good and excellent.

- 43% of state agencies rated the quality of their relationship with the regional/Washington office as excellent or very good; 86% rated the quality of the relationship as between good and excellent.
- 47% of state agencies rated the quality of their relationship with individual refuges as excellent or very good; 88% rated the quality of the relationship as between good and excellent.

Benefits, Costs, and Challenges of Partnerships: The use of partnerships and volunteers has been actively promoted in the Refuge System and has been a cost effective way to increase the level of habitat management and visitor services the NWRS is able to provide. The management and care of these programs, however, require staff time, and a large portion of refuge managers (80%) indicate that they do not have adequate staff to manage volunteers and partners. While the growth in the use of partnerships and volunteers has been highly effective, it is likely that these levels are at or near capacity due to limited refuge funding and staff.

The Conservation in Action Summit: The Conservation in Action Summit, held in 2004, was a Refuge System-sponsored event that brought together 250 refuge employees and partners to identify future conservation priorities and areas in which to strengthen partnership. While the Conservation in Action Summit was a significant effort to involve partners in determining Refuge System priorities, many of the recommendations identified at the summit appear to have not been implemented, particularly in areas related to environmental education and research, the use of the Habitat Goals Process to develop stronger eco- regional plans, and implementation of a system- wide bird and habitat monitoring program. In part, these recommendations were not implemented because they coincided with a decline in the Refuge System's budget. In addition, it does not appear there was a sufficiently structured follow-up process to develop work plans to implement the Summit's priorities.

RECOMMENDATIONS

Volunteers: Volunteers make a tremendous contribution to the Refuge System and are a cost- effective way to accomplish many tasks. The Refuge System should endeavor to continue its efforts to develop stronger partnerships across the system. This would include continuing to provide grants for volunteer development and providing leadership training to volunteer groups.

If the Refuge System wants to further increase the use and contributions of volunteers, it may need to commit additional staff to the task. The Refuge System should look to develop dedicated positions for managing volunteer and partner contributions, particularly where geography and interest provide good possibilities for doing so. In addition, the Refuge System should consider identifying best practices, both programmatically and managerially, for managing the use of volunteers. One way to proceed might be to develop profiles identifying which refuges have the best prospects for attracting volunteers and partnerships, e.g. those near population centers of a reasonable size and with a modest or higher level of visitation, and then explore the possibilities of assigning dedicated staff for partnership and volunteer development at the refuges where such an approach is likely to produce the most benefits.

—*Ryan Hagerty*- FWS Volunteer Holds Prothonatary Warbler for Young Girl at Bird Banding Event.

Friends Groups: The Refuge System should continue to invest in the Friends program through the actions that have nurtured the existing groups. Grant programs are a strong mechanism for strengthening these programs, as are national and regional workshops that serve as forums for sharing best practices.

SOG 7: Protect Resources and Visitors through Law Enforcement

Performance Rating: Performance Rating

There are a number of positive developments to point to when considering the state of law enforcement at NWRS: law enforcement training is sound and is improving; Refuge law enforcement officers are judged by their supervisors and colleagues to have the skills, abilities, and commitment necessary to meet their responsibilities; and policies to guide and manage the law enforcement program are beginning to be developed and disseminated. This good news, however, is offset by a very serious shortcoming: there is a critical lack of law enforcement coverage at most of the system's field stations (more than 70% of Refuge managers feel law enforcement coverage is insufficient at the refuge they manage). In addition, it appears that this problem, which began with law enforcement reforms in 2003, is probably trending in the wrong direction, i.e. coverage may fall to even lower levels in the near future. This is a particularly troubling issue given that serious crime on refuges has increased in recent years and will likely continue to do so as more refuges find themselves located near population centers. Until there is a significant increase in the number of trained law enforcement officers deployed at NWRS field stations performance against this SOG will remain ineffective.

Conclusions

Staffing levels: Low staffing levels are leading to a substantial and critical lack of law enforcement coverage and capability at many refuges across the system. At many refuges, law enforcement coverage is insufficient to ensure the protection of resources and the safety of visitors and refuge staff. A substantial majority of refuge managers (over 70%) feel visitor safety and law enforcement performance have declined in recent years. The issue of public safety is of particular concern given that only seven of the refuge managers from 50 high visitation refuges (with annual visitation in excess of 250,000) who responded to the MSI survey indicated that law enforcement coverage is sufficient on their refuge.

Law enforcement staffing levels are far short of both the recommendations of the International Association of Chiefs of Police (IACP) deployment model and more conservative NWRS internal law enforcement needs assessments.

"The Refuge complex is made up of five different geographically located units covering 115,000 acres. The Refuge has only one full time law enforcement officer to cover this land base which is totally inadequate. Additionally, most new officers are trained toward the non-resource protection enforcement with few ambassador skills. The abolishment of collateral officers has significantly hampered the Refuge's ability to protect resources and the public." *...quote from MSI Refuge Manager's Survey.*

In addition, as LE budgets have leveled off in recent years, the move towards an improved staffing situation has stalled. It is highly unlikely that any meaningful progress towards improving the Refuge System's law enforcement capability under current and expected budget allocation levels.

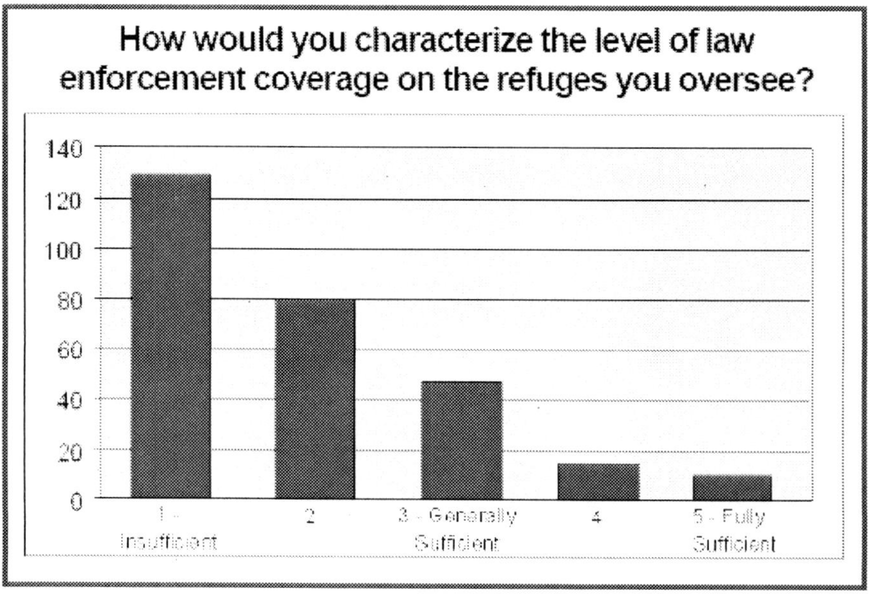

Figure 5. Results of the 2007 MSI Refuge Managers survey on the adequacy of current law enforcement coverage.

We have no law enforcement officers on any of the 3 refuges. We rely on officers from other stations to help us. The LE situation is embarrassing. We are not protecting our natural resources or our visitors. It all started when the Service drastically reduced the use of dual function officers. The full time officers are good officers, we just need a lot more officers to protect our refuges." *...quote from MSI Refuge Manager's Survey*

Shift to Dedicated Law Enforcement Officers: As noted by refuge managers, the move away from collateral duty officers to develop a more professional law enforcement capability composed of full-time officers has not been implemented to the extent that is required. The number of full time officers in the field has increased slowly since reforms began in 2003, but the increase has been insufficient to fill the gaps left by "decommissioned" collateral duty officers. Though the various reviews and reports of the law enforcement function at DOI and the Refuge System made clear the need to move away from the use of collateral duty officers, the lack of substantial increases in the deployment of full time officers has left refuge managers feeling that the "old" system of law enforcement deployment was dismantled and that the "new" system was never implemented.

Though all regions have too few Law Enforcement (LE) staff to provide adequate law enforcement coverage to their field stations, Region 3 is particularly hard hit. The region has only 17% of the law enforcement officers recommended by the IACP model—by far the largest regional gap between recommended and actual LE staff levels. Not surprisingly, only 11% of Region 3 refuge managers indicate that LE coverage is adequate on their refuge.

Training: The training of law enforcement officers appears to provide officers with the skills and knowledge necessary to meet their LE responsibilities in an appropriate and effective fashion. However, it is not yet clear if the training of NWRS LE officer supervisors will adequately prepare the supervisors for their role.

Role of Zone Officers: Though additional data is required to confirm initial anecdotal findings, nearly three years after relevant reforms were initiated it appears zone officers are not fully utilized as law enforcement resources, particularly by field stations. This is particularly troubling given staffing shortages at the field station level. It is also an issue because zone officers were intended to address—in a compromise fashion—the OIG's call for centralized lines of command.

RECOMMENDATIONS

Increase the Number of Full-time Law Enforcement Officers: Increase the number of commissioned law enforcement officers deployed in the field in order to address the critical lack of law enforcement coverage across the Refuge System. The severe gap in staffing can only be addressed by hiring additional full-time law enforcement officers—moving from current levels of around 200 full-time officers to at least 400 full- time officers. Refuge Managers clearly see this as a critical issue, noting that a continuing increase in the risks to public safety, staff safety, and resource protection are unacceptable. Implementation of this recommendation will require substantial resources, but an acceptable improvement in law enforcement coverage is of fundamental importance to the on-going effectiveness of the Refuge System.

—*John and Karen Hollingsworth*- Refuge law enforcement officers have traditionally been responsible for monitoring compliance with wildlife laws but, in recent times, there has been a marked increase in serious crime on refuges, including vandalism, armed robbery and illegal drug activity.

Prioritize the Deployment of New Officers to Address the Most pressing Needs: Any new law enforcement officers should be deployed to those regions and refuges at highest risk. For example, Region 3 may warrant first consideration, and special attention should be paid to high visitation refuges.

Retain Existing Dual-Function Officers as Possible and Necessary: Do not push to replace existing dual-function officers as dual-function officers will continue to play a critical role in refuge law enforcement until such time that more full-time LE officers can be hired; however, this is a stop-gap measure, at best, because the nature of law enforcement and the increasing complexity of crime on refuges requires that the Refuge System move to an adequate force of full-time officers as quickly as possible.

Explore the Use of Interim Measures to Address the Most Severe Gaps in Law Enforcement Coverage: The parameters of the evaluation did not allow for thorough examination of all potential options for interim measures. However, the evaluation team directly observed or discussed with Refuge System staff several ideas, including, for example:

Utilize Short-Term Deployments to "Share" Law Enforcement Officers During Periods of Low Law Enforcement Activity: Many Refuges experience seasonality with regard to law enforcement requirements, i.e., there will be peak months during which law enforcement coverage is critical, but also months during which LE demands are greatly reduced. The Refuge System should explore the possibility of deploying officers, on a short term basis, from refuges with low seasonal LE demand to refuges with high seasonal LE demand. Similar to the preceding recommendation, this should be seen only as a near term, stop gap measure.

- Expand Partnerships with State Law Enforcement Staff Wherever Practical: State law enforcement officers currently work with Refuge System officers to support—or fully manage—select public use activities on refuge lands (e.g. managing hunts). Opportunities for such partnerships should be fully explored and, wherever practical, put into place.
- Close Refuges to the Public: In order to focus scarce law enforcement resources on the highest priority needs, close selected refuges in each region until such time that adequate law enforcement capability exists. The Refuge System should identify criteria to facilitate the process for identifying refuges for closure perhaps using the process Region 5 utilized for similar purposes as a roadmap.

Improve Recruitment and Retention of Law Enforcement Officers: The current pay grades and opportunities for advancement available to Refuge System law enforcement officers provide minimal incentives for individuals to pursue—or remain in— law enforcement positions within the Refuge System. Better pay grades and opportunities available to law enforcement officers at other USG land management agencies put the Refuge System at a distinct disadvantage in its efforts to recruit and retain law enforcement personnel. The Refuge System should modify the job descriptions for law enforcement officers to allow for higher GS pay grades and/or more attractive mid-career opportunities (including lateral moves out of law enforcement to other positions in the Refuge System). In addition, the recruitment process should be reviewed with an eye to developing new recruitment vehicles or better utilizing existing Refuge System programs such as the Student Temporary Employment Program (STEP) and the Student Career Experience Program (SCEP).

Review the Role and Deployment of Law Enforcement Zone Officers: At a minimum, Zone Officers are underutilized in some NWRS regions. This situation may be a result of: a) the number of zone officers deployed is based on the assumption that there would be a much greater contingent of full-time officers to oversee than is currently the case or; b) a lack of operational clarity regarding the role and responsibilities of zone officers, given the lack of line authority between zone officers and the refuge-based officers they "oversee." Consideration should be given to adjusting the responsibilities and tasks currently defined for Zone Officers and perhaps deploying fewer. The savings in resources from a reduction in the current number of Zone Officers could be used to hire additional full-time officers. In addition, an assessment of the role of Zone Officers may identify useful "best practices" that could be standardized as required tasks across the system, e.g. refuge level "law enforcement assessments" that identify important law enforcement risks and opportunities

SOG 8: Provide Infrastructure and Equipment Adequate to Support Mission and Maintained in Good Condition

Performance Rating: Effective

In the mid-1990s, the maintenance of the Refuge System's infrastructure and equipment was a critical concern, and the maintenance budget increased dramatically - from $21 million in 1996 million to $91.5 million in 2004 (a 336% increase over eight years). The availability of increased funds over the past seven or eight years has allowed the Refuge System to effectively address preventive maintenance requirements, target the most urgent deferred maintenance projects, and selectively add new facilities. Largely as a result of these additional resources, the majority of refuge managers do not currently view maintenance concerns as a constraint to achieving their refuge's purpose: 75% of refuge managers indicate that the condition of the facilities on their refuge is at least sufficient to support the purposes of their respective refuges (2007 MSI Refuge Managers Survey). Subsequent to 2004, however, maintenance funding dipped substantially – a decline of 30% from 2004 to 2007. It is important to note that if the recent backsliding in maintenance funding is not reversed infrastructure maintenance will soon become a critical problem again.

An important NWRS initiative in this area over the past several years has been the implementation of the Service Asset Maintenance Management System (SAMMS), a maintenance management software system intended to provide better information to guide decision-making at the national, regional, and refuge levels. SAMMS is starting to provide improved information at the national level, but it has not been well-received at the field level. SAMMS is viewed as one of the major sources of the recent increase in administrative burden being shouldered by field station staff, and that burden has not been offset by any perceived value—80% of refuge managers feel their ability to manage maintenance needs is about the same or has decreased with the introduction of SAMMs. NWRS headquarters, recognizing the issue, has continually modified SAMMS to reduce the burden of data input and management at the field level. It is still unclear, however, whether SAMMS will ever be of substantial value to refuge managers.

CONCLUSIONS

Maintenance Funding: The budget situation for maintenance—both annual and deferred maintenance—has improved dramatically as compared to ten years ago. The increase in maintenance funds, adjusted for inflation, is well over 200% over the recent ten-year time span. Since FY 2004, however, maintenance budgets have declined, particularly the deferred maintenance budget, which has decreased in real terms by 30% over the past three years. The recent budget declines are especially worrisome because the Refuge System's asset inventory continues to grow every year from both new construction and acquisitions. If the recent declines in maintenance funding are not reversed, the backlog issue will rapidly become a problem that will be difficult to resolve.

Maintenance of Critical Assets: The most important refuge assets—those most necessary to the achievement of refuge conservation and public use objectives— are generally well maintained. These "high" Asset Priority Index (API) assets generally have a Facility Condition Index (FCI) of less than 10, indicating they are in "fair" condition. In addition, 75% of refuge managers feel that the assets most critical to their refuge's mission and purpose are maintained in a condition adequate to support and achieve those goals. An important caveat to this conclusion is the fact that a substantial minority of refuge managers (40%) believe their refuges require new facilities if they are to meet their purpose and objectives.

Serving as a counterpoint to the preceding conclusion, a number of facility categories rate as being in poor condition. Some of these asset groups, though not of the highest level of importance as defined by the API, still impact substantially on the ability of the refuge to be effective in meeting its conservation and public use objectives, e.g. refuge offices and contact stations.

The process of selecting deferred maintenance (DM) projects for funding does not follow specific system-wide standard procedures. The identification of DM projects to receive funding, conducted at the regional level, follows only general guidelines offered in the FWS Asset Management Plan and the SAMMS Business Rules. These guidelines are very generic and essentially allow any factor to determine selection. Currently, with such a tight DM budget (as compared to the backlog), it is likely that all funded DM projects are of high priority. However, without a national or system-wide process for identifying the highest priority DM projects it is possible that increases in DM funding will result in lower priority DM projects being funded, when considered on a national level.

Maintenance Staffing: Maintenance staffs are stretched thin, maintaining on a per capita basis, approximately 60 facility assets and 20 vehicles while also being tasked with non-maintenance responsibilities. Looking ahead, the asset inventory of the system will almost certainly continue to increase, particularly with regards to constructed assets. In addition, the tasks necessary to manage the asset base have increased in breadth and complexity, e.g. managing the SAMMS process, and will likely continue to expand in the years to come. It is not clear that the skills and competencies necessary to effectively carry out some of these emerging maintenance-related tasks are widely evident in the current complement of maintenance staff.

SAMMS: It is still too early to draw a clear conclusion on the value of SAMMS. SAMMS is able to receive and manage a broad spectrum of asset-related data, and it should facilitate integration of asset information with Department-wide budget and information systems currently in development. However, it is also complex, generally not user- friendly, and represents a substantial burden to many field staff that already have too much to do and too little time to do it. To the credit of the central managers of SAMMS, many adjustments aimed at reducing the burden and complexity of SAMMS in the field have been put in place. Still, if SAM MS cannot be shown to have value as an information source (e.g. producing reports that will inform management decisions at all levels of the Refuge System) it is unlikely that it will ever be perceived by the field as anything more than an administrative burden.

RECOMMENDATIONS

Maintenance Funding: Restore maintenance funding to levels in line with the FY 2004 budget. In general terms, the infrastructure and facilities that are most critical to the achievement of the mission and goals of the Refuge System are currently well maintained. However, if funding declines are not reversed, this will soon no longer be true. Related to the first recommendation, fully recognize and budget for the maintenance costs related to new construction and acquisitions. The Refuge System's asset inventory continues to expand rapidly—a fact which is obvious, but which also seems to be ignored when developing maintenance budgets.

Establish an objective, transparent, and standard process for identifying priority deferred maintenance projects that are to receive funding. The idea is to fund the projects that have the greatest value to conservation and public use objectives when thinking about the Refuge System as a whole.

Maintenance Staffing: Managing and maintaining the asset base of the Refuge System is an increasingly complex undertaking, moving beyond the typical current skill set of maintenance staff (and others, such as biologists and refuge managers who are often tasked with the supporting the maintenance function in the absence of sufficient maintenance staff). The Refuge System should examine a move towards a different staffing approach that would utilize a group of "asset management specialists" to meet these more specialized asset management responsibilities (e.g. SAMMS, etc.). This should not only result in more effective and efficient asset management, but will also allow maintenance staff (and others including biologists and refuge managers) to focus on tasks more directly aligned with their skills and competencies. In this light, the experience of Region 4—that is, the deployment of a small group of specialists to handle the majority of SAMMS requirements for all refuge units in the region— may prove instructive.

SAMMS: As the development and refinement of SAMMS continues, focus on developing its ability to produce useful reports or analysis for managers at all levels of NWRS. Start small and simple—there is no need to create a complex series of SAMMS- produced reports. By clearly demonstrating the value of SAMMS to refuge staff there is a greater likelihood that managers will be willing to put time into utilizing and developing SAMMS as a field-level management tool. In addition, consideration should be given to increasing the user-friendliness of SAMM's computer interface, so that the system becomes more intuitive and does not require extensive training to master or manipulate. If, after some reasonable period of time, SAMMS cannot demonstrate its value to the field then the system it should be re-designed to be less burdensome (or dropped in favor of an alternative system/process).

SOG 9: Complete Quality and Useful Comprehensive Conservation Plans on Schedule and with Full Engagement of Partners

Performance Rating: Effective

The NWRS is required to complete CCPs for 554 refuges by 2012 and to date has completed somewhat over 200. Although the pace of CCP completion has accelerated significantly over the past few years, the Refuge System is slightly behind schedule in terms of meeting its CCP completion target. This is mainly because a few regions are not on pace and may require additional support or additional time. In April 2007, the Refuge System began implementing the 2012 Plan, An Action Plan to Meet Our Legislative Mandate, which lays out a series of actions intended to ensure that all required CCPs are completed by 2012.

Overall, refuge managers have found CCPs to be a useful tool for clarifying objectives, guiding habitat management decisions, and clarifying public use decisions. The policy to develop CCPs for all refuges has improved the Refuge System's interaction with stakeholders and has helped to create a more professional approach to planning and management. In response to an MSI survey, 94% of those state agencies that responded agreed or strongly agreed that they had been provided an opportunity to meaningfully participate in the CCP process; and 95% of state agencies agreed or strongly agreed that their participation in the CCP process had improved communication and coordination between their agencies and the Refuge System.

Conclusions

CCP Completion Rate: Throughout the Refuge System, CCP completion activity has increased significantly over the past five years and in particular over the past two years. Two hundred and five refuges are now covered by completed

CCPs—about 37% of the system. About half of NWRS regions appear on-track to completing CCPs by 2012. For several regions, most notably 1, 2, 4 and CNO, the current pace of CCP completion appears insufficient to meet the NWRS' overall goal. Region 4 must still complete CCPs for 79 refuge units over the coming five years. However, the Refuge System has recently revisited the rate of CCP completion and has developed a 2012 Plan that lays out actions and a schedule that should enable all required CCPs to be completed by 2012.

While it may be possible to complete all required CCPs by 2012, the accelerated rate of completion required may also raise issues as to whether this will lead to trade-offs in terms of the level of detail included or in the quality of the plan. Currently there is a great deal of variance between the level of detail contained in CCPs—some are hundreds of pages long and contain enough information to serve as detailed Habitat Management Plans and Visitor Service Plans, whereas others are shorter, less detailed documents that will require the completion of separate step-down plans.

CCP Usefulness: CCPs and the CCP development process have proved to be very useful tools for refuge managers. This is confirmed by the MSI Refuge Managers Survey in which 72% of managers characterized CCPs as being useful or extremely useful. In particular, refuge managers cited CCPs as being useful for the following: setting goals and objectives; development of annual work plans and activities; managing habitat; and determining appropriate public uses of the refuge.

CCP Implementation: While managers indicate that CCPs are being used to guide goal setting and work plan development, it is also clear that more often than not CCP designs exceed the level of funding that is likely to be able to implement the priorities identified—92% of refuge managers indicated that funds are less than sufficient to implement CCP priorities.

As a general rule, strategic plans are intended to identify future priorities that can be initiated, scaled-up, or cut back as circumstances and funding levels change. In the case of CCPs, it would appear that many plans have not been used to prioritize proposed activities in relation to existing or likely budgets. There is some debate (or lack of clarity) in the Refuge System as to whether the CCP is supposed to serve as a vision document, which lists an ideal set of tasks and objectives to be accomplished irrespective of available or likely funding, or whether the CCP is supposed to be a more practical planning document to prioritize implementation activity over the coming fifteen years.

CCP Quality and Consistency: CCPs undergo a multi-level review process within Regional Offices. In general, these reviews have been sufficient to produce CCPs that refuge managers feel are of adequate quality. A peer review process undertaken by stakeholders and outside experts, including managers from other refuges, is undertaken in some regions.

State Fish and Game Agency Participation: State agencies gave the Refuge System very high ratings in terms of their participation in the CCP process.

- 94% of state agencies agreed or strongly agreed that they had been provided an opportunity to meaningfully participate in the CCP process; and
- 95% of state agencies agreed or strongly agreed that their participation in the CCP process had improved their communication and coordination with the Refuge System.

RECOMMENDATIONS

CCP Completion Rate: Additional attention and resources will be required to complete CCPs on time in several regions, including in regions 1, 2, 4, CNO and 7. Under the current pace of CCP completion; it is likely that the Refuge System overall should be able to complete upwards of 90% of required CCPs by 2012. The Refuge System's recently completed 2012 Plan: Meeting Our Legislative Mandate has developed a schedule that regions have approved and which, if followed, will ensure that all required CCPs are completed on schedule.

CCP Usefulness/Implementation: The NWRS should provide better guidance for what level of budget resource should be assumed when producing CCPs. Since 92% of refuge managers say that budgets are insufficient to implement key priorities, it is clear that there is a gap between the budgets required to implement CCPs and the level of funding available.

It is suggested that refuge managers prioritize the activities in their CCPs so that in the annual work planning process it is clear which activities will be implemented within the parameters of limited funding. For example, managers could note the objectives and initiatives that could be undertaken under current funding scenarios and those activities that would be priorities if funding were to increase (for example, provide a general description of alternative CCP implementation scenarios which factor in inflation costs and are based on existing budgets and a budget increase, for example a 10- 15% increase). This will help CCPs to serve as a mechanism for prioritizing refuge activities—i.e., for explicitly identifying what can and cannot be implemented within differing resource ranges (and to avoid CCPs becoming laundry lists of activities that are not prioritized and which will never be fully implemented because the cost substantially exceeds likely resources).

CCP Quality and Consistency: There should be a review of CCP content and guidance developed on the appropriate level of detail to be contained in a CCP, in contrast to what should be included in step- down habitat plans. Better guidance on review processes could also be developed to ensure that all CCPs meet high quality standards. In the Fire Program, there is significant cross-refuge collaboration in the development of Fire Management Plans, which serves as a peer review process and mechanism for sharing knowledge and information and helps participating personnel gain skills that they will use in developing their own plans. A similar process might be useful to institutionalize for CCP development. In general, a lot of emphasis has been placed on developing CCPs processes. At this point, it might be worthwhile to increase the standardization of content and review processes, including reviewing the specificity of objectives and ensuring these are linked to monitoring and inventory plans.

SOG 10: Strategically Grow the Refuge System

Performance Rating: Ineffective

This objective was rated ineffective for a number of reasons, including: the rate at which land has been added to the NWRS has declined significantly over the past five years; land purchased by the Refuge System often does not match the priorities identified by the NWRS' Land Acquisition Priority System (LAPS), especially over the past few years; and the current DOI-managed land appraisal process that the NWRS uses is ineffective and cannot be relied upon to produce timely or accurate appraisals and this causes available land deals to be lost. The NWRS does not have a common system-wide approach to landscape-level planning that can drive real estate acquisition decisions; however, several select field stations have developed sophisticated state-of-the-art biological-based planning systems that can serve as models, e.g. HAPET. It is perhaps also worth noting that real estate acquisition is an inherently political process and that ultimately land acquisition decisions are made by the Congress, which does not always base its decisions on the priorities of the Refuge System as defined by LAPS. It is recommended that the Refuge System develop a land acquisition policy that incorporates the principles contained in the Strategic Habitat Conservation Initiative.

Conclusions

Land Acquisition: The rate at which the Refuge System has acquired new land has slowed markedly over the past five years and the number of acres added to the NWRS between 2002 and 2006 has declined significantly as compared to the preceding five-year period. This has been mainly due to a decrease in allocations in Congressional Appropriations, from $121.1 million in FY01 to $24 million in FY06—a decline of about 80%. However, the Refuge System also significantly decreased its requests to add new lands to the system: in FY2008 the Refuge System only requested to purchase two new lands as compared to a request for 21 purchases in FY2005. Between 1996 and 2006, the NWRS grew by 2,851,469 acres, or by about 3.17%; between 2001 and 2006 the amount of land in the system increased by about 0.8%.

Over the past several years, the NWRS decreased the amount of land it has requested to purchase. The number of requested properties to be added to the Refuge System was 53 in FY03 and has declined every year since. In FY08 only two properties were requested for addition to the NWRS—despite the FY07 LAPS priority list being composed of 128 available properties. (Note: the LAPS system does not currently identify or rate opportunities for adding new refuges but only prioritizes projects that have an acquisition component.) This is likely due to land acquisition perhaps being somewhat out of favor with the current administration and also due to declining budgets—which reduce funds for acquisitions but also make it more difficult to purchase new land in light of the difficulty of maintaining existing lands with declining resource levels. In addition, over the past several years (since the adoption of the ASD appraisal process), the Refuge System's real estate staff has declined by about half.

Land Appraisal Process: The current DOI-managed land appraisal process is ineffective and represents a step backward in the NWRS' ability to purchase land and easements from willing sellers. In some cases, the process is so slow that opportunities to purchase land and easements are lost. Fifty-eight percent of survey respondents in the Refuge Manager's Survey said that the current appraisal process "decreases our ability to acquire land."

Land Acquisition Policy/Strategy: The Refuge System does not have an existing Land Acquisition Policy or an effective land acquisition strategy. As recently as FY05, there was a high correlation between the lands that NWRS requested to purchase and the priorities listed on the LAPS system. However, in recent years, there has been a significant divergence between land purchase requests and the priorities indicated by its own system—to the point that the NWRS no longer appears to be using a transparent criteria-based system to prioritize land purchases. It is unclear if the current divergence between acquisition decisions and LAPS priorities is due to deficiencies in LAPS or because of other factors in the decision- making process. An additional shortcoming of LAPS is that it only deals with additions of land for existing refuges and does not consider the need for adding new refuges.

Landscape-level Planning: The NWRS does not have a common system-wide approach for landscape-level planning that can be used to identify land acquisition priorities, although excellent models were found at select refuges. In order to identify and prioritize the establishment of new refuges, or to compare the relative value of competing opportunities to add land to existing refuges, there needs to be an overall landscape- scale strategy upon which to base decisions. There are various examples of the use of such systems within the Refuge System, most notably the HAPET system at use at the Fergus Falls WMD, but such approaches do not appear to be the norm.

Table 6.

From the 2007 Refuge Manager's Survey: How does the current Department of Interior real estate appraisal process affect your ability to acquire additional refuge land from willing sellers (in comparison to when the process was directly managed by the NWRS)?	
Significantly increases our ability to acquire land	1%
Somewhat increases our ability to acquire land	1%
No significant difference	14%
Somewhat decreases our ability to acquire land	26%
Significantly decreases our ability to acquire land	58%
No. of responses considered: 209; 83 respondents answered "not sure" and there answers were not considered in the above percentages	

Recommendations

Land acquisition is among the FWS programs most affected by politics. Nevertheless, there are a number of steps the NWRS can take to improve its land acquisition strategies and processes.

Land Acquisition Policy/Strategy: The Refuge System should develop a Land Acquisition Policy and a corresponding strategy to guide expansion of the system. It is recommended that the Refuge System's new land acquisition policy be developed to be consistent with the Strategic Habitat Conservation Initiative framework.

Land Appraisal Process: The Refuge System should engage in a discussion with the Department of Interior to enable it to cease using the services of ASD and return to its former system of direct management of land appraisal. This would save time and money and would enable the NWRS to increase the effectiveness of its land acquisition program. This would also result in less land purchase opportunities being lost to inefficient and ineffective administrative processes. The appraisal process should include the flexibility to value land based on the inclusion of common valuation considerations used by the private sector, e.g. increased value due to the proximity to designated conservation lands.

Land Acquisition Strategy/Landscape-level Planning: In the near term, the NWRS should revisit its decision-making process and its LAPS process to increase transparency and ensure land acquisition decisions are criteria-based. There will likely always be cases when real estate purchases make sense even though they are not LAPS priorities, but the NWRS should endeavor to have a process that makes this the exception rather than the rule. Possibly, LAPS could be expanded to be a more all- encompassing land acquisition prioritization system by also ranking opportunities to purchase lands for new refuges.

Over the medium to longer term, the NWRS should engage with FWS to develop a planning system that identifies geo-spatial species priorities and uses this information as a basis for identifying habitat needs and objectives Such a system would require significant input from other FWS programs primarily from those parts of FWS responsible for monitoring species and setting species goals, including the Migratory Birds, Endangered Species, and Fisheries Programs, and could possibly be built in collaboration with USGS. Such a system would enable the Refuge System to better plan management actions for existing habitat and better prioritize the acquisition of new habitat, as per species goals. The absence of an FWS overall species-based and landscape-scale planning system limits the Refuge System's ability to prioritize new land acquisition opportunities.

In addition, a better landscape-level planning system could help FWS increase its ability to work effectively in combination with others to build more holistic conservation landscapes (as the amount of funding FWS has to acquire land is relatively modest in comparison to other available sources of conservation financing). A useful entry point for re-visiting the Refuge System's approach to land acquisition would be to develop/finalize a land acquisition policy.

SOG 11: Reduce Wildfire Risks and Improve Habitats – Reduce the Threat, Risk, and Adverse Effects of Unwanted Wildland Fires by Reducing Hazardous Fuels, Restoring and Maintaining Fire Adapted Ecosystems in Lower Condition Classes, and Improving Fire Prevention and Suppression Capabilities of The Service and of Neighboring Rural and Volunteer Fire Departments

Performance Rating: Effective

This objective is rated "Effective" as a result of the systematic planning and execution by which the NWRS utilizes prescribed fire to improve wildlife habitat and reduce fuels loads and also for its ability to fight and suppress wildfires. At refuges with proper staffing and adequate budgets, this program is "Highly Effective," but many refuge units do not have proper staffing and adequate budgets.

Conclusions

Where refuges have the qualified staff and budget, the high level of planning, training, and coordination results in application of prescription fire to improve and maintain habitats, reduce fuels loading, and suppress unwanted wildfire. From surveys and interviews, it appears that approximately one-half of the NWRS has the resources it needs – both budget and personnel – to use fire as a habitat management tool. For other units, issues of staffing, available budget, the growing percentage of wildland-urban interface lands (WUI), and the location of a refuge relative to other fire resources, often impairs a refuge's ability to promote prescription fire while proactively addressing fuels availability and effective wildfire suppression.

—*USFWS*- Refuge Fire Management Specialist conducts a prescribed burn to improve wildlife habitat.

Fire Planning: The Refuge System has an excellent system for fire planning. Refuges with "burnable vegetation" must have an approved Fire Management Plan (FMP). These are long-term plans that must be updated annually. Virtually all refuges have completed FMPs. The practice of having neighboring refuge staffs participate in the development and review of FMPs is a practical process for ensuring quality control through peer review and for building the experience and skills of fire management officers and other staff.

Budget Structure: The funds available for fire management are currently classified into fire suppression and fuels reduction categories and there are no dedicated funds available for prescription burns that are primarily undertaken for the purpose of habitat improvement. Thus, refuge managers often must creatively try and use funds designated for fuels reduction when the primary purpose of a prescription fire is habitat improvement. This complicates the ability of managers to access funds for habitat improvement burns and likely prevents fire from being used for habitat management to the extent that would be desirable. As WUI issues increase there will be added pressure to continue to use funding for fire suppression and not habitat management.

Increased Cost of Wildfire Management: Similar to the National Forest System and other public lands, the NWRS is experiencing the increased cost, complexity and frequency of fire management in the WUI. Increased staffing and costs related to suppression efforts will pose greater challenges to the NWRS than they have with other federal land management agencies. Increased resource calls on staff and budget to fight fire in the WUI can and will impact the ability of the NWRS to use prescription fire and fuels reduction proactively to benefit habitat condition and reduce wildfire risk. The accelerating population growth in the western United States, especially in the WUI, coupled with multiple-year drought conditions, are factors in this equation.

The high level of cooperation and coordination exhibited by the Fire Program across refuge, agency and state boundaries is requisite for an effective fire program, but it also provides a Best Management Practice for possible application to the rest of the NWRS.

Growth of residential development in areas adjacent to refuges will continue. For refuges facing increased difficulty conducting prescribed burns and increased risk of wildfire in the WUI, active involvement with surrounding communities and regional planning agencies is important.

RECOMMENDATIONS

Budgetary Restrictions: As guidance is currently structured, it is difficult for some refuge managers to access funds for using fire to manage habitat. This may because of a misunderstanding in terms of how fire funds can be used, or there may be a need to better communicate to regional staff the flexibility that is possible in using such funds. In either case, there is a need to address this constraint so that refuge managers have greater flexibility to program funds for using fire to improve habitat. In addition, a separate budget category should be established and funded for using prescription fire for habitat improvement (as distinct from fuel suppression).

Plan for the Increased Frequency and Complexity of Fire in the Wildland Urban Interface: Growth of residential development in the WUI will increasingly impact the ability of the NWRS to manage habitat. This issue clearly illustrates the need for refuges to engage in the larger community, not only in a biological landscape sense, but also to influence regional growth and development. The Refuge System, and other federal agencies, has been able to have some influence on development practices in wildland areas adjacent to refuges through applying "communities at risk" designations. This tool should be reviewed, and possibly more broadly applied, as a way to influence development in fire prone areas adjacent to refuges.

SOG 12: Promote and Enhance Organizational Excellence

Performance Rating: Partially Effective

The Refuge System has introduced a number of new management and planning systems over the past several years, including a medium-term strategic plan, activity- based costing, RAPP work planning and reporting systems, and refuge-level comprehensive conservation planning. The Refuge System is also currently undertaking a Workforce Planning exercise to help better balance personnel and operational expenditures and to prioritize staffing and programs in consideration of declining budgets. The RAPP system has enabled the NWRS to better track and report on national-level accomplishments and the budget rebalancing exercise will, over time, provide managers greater ability to address local priorities. There are, however, several aspects of the Refuge System's overall management system that could benefit from additional attention, including: increasing policy implementation consistency across regions; standardizing business management processes across regions, such as annual work planning and budgeting; and building systems to better analyze performance, share best practices and better connect communities of professionals, such as biologists and visitor service staff.

Conclusions

The NWRS has made a significant progress in recent years in several key management areas. These accomplishments include:

- The completion of a medium-term strategy, which defines the NWRS priorities and guides its programs.

- The completion of a significant number of refuge Comprehensive Conservation Plans, which set refuge objectives and engage partners and stakeholders in planning how objectives will be accomplished.
- The development of the Refuge Annual Performance Plan planning and reporting system, which has enabled the NWRS to track and report national-level accomplishments.
- Instituting a Workforce Planning process that will restructure refuge expenditures so that all refuges will have an appropriate balance between personnel costs and operational funds (which has been variously defined as an 80/20% or a 75/25% balance). This will increase the ability of refuges to respond to and manage annual needs, although it will also leave some regions with fewer staff.

These improvements have definitely strengthened the Refuge System's management operations. Management improvement, however, is a continual process. While it is certainly the case that important progress has been made in recent years there is also more that needs to be done. In particular, attention needs to be given to the following areas:

- There is a need to increase policy and program consistency across regions. The Refuge System often does not operate as a system but rather operates as a series of eight parallel regional efforts. Some advantages of operating more as a more coherent system would include better cost controls, not having to reinvent practices in each region, and learning from and applying best practices across regions so that minimum quality standards can be achieved in different practice areas.
- There is need to establish a refuge budget allocation process that is consistent across regions and is able to deliver budgets to refuge managers in a timely manner and provides the flexibility refuges need to respond to changing situations and challenges. While some of the delay in provision of refuge budgets can be attributed to delays in congressional appropriations, there are opportunities to improve the efficiency of the Refuge System's budget allocation process.
- It would be beneficial to create processes to analyze and take stock of experience and to create a system that will effectively identify and disseminate best practices and better connect communities of professionals.

The advantage of having greater consistency in management approaches and practices is twofold: 1) to ensure that all managers are using the most effective and efficient approach possible; and 2) to ensure that all managers can apply standards to enable their operations to achieve a standard level of quality, which includes learning from and applying the lessons from similar experience gained elsewhere in the system.

A series of decisions and actions will be required to create a more effective NWRS management system as there are a number of inter-related issues that could benefit from attention. These issues include:

- The lack of centralized management authority, which has resulted in variances in how regions operate;
- A general reluctance to analyze the comparative effectiveness of differing approaches and then more widely adopt the approaches that show a comparative advantage;
- The lack of standard business operating procedures; and
- A strategic planning and reporting process that is cumbersome and more oriented toward Washington reporting needs than to the needs of refuge managers.

These and associated issues are discussed in the conclusions that follow.

Business Management Processes: The Refuge System has not developed and implemented a consistent set of standard business operating practices, and this has led to a great degree of variance within the system in terms of how work is planned, budgeted, and assessed. At the refuge level, there is no standard or required annual work planning process, there is no standard budgeting process, and assessments/ evaluations of refuge effectiveness are not generally being conducted. Despite the lack of standard business processes, this assessment did find that the Refuge System has a highly motivated and highly experienced cadre of refuge managers. Thus, while some and/or many refuges are well-managed, there also appear to be many refuges that appear to fall short of achieving desired standards. Furthermore, the lack of standard approaches makes it difficult to assess how the system is performing overall or to introduce improvements system-wide to help all refuges reach a level of minimum acceptable quality.

Comments on specific business practices follow.

- *NWRS Strategy*: The current strategic plan does not contain a causality logic model, whereby one can identify how long-term objectives will be achieved and whether the linkage between program objectives and program activities is successful. This makes it difficult to determine if the overall strategy is meeting with success or to determine weaknesses in the strategy's logic, assumptions, or implementation. A logic model helps to ensure that all activities, existing or anticipated, contribute directly to the achievement of outcomes and results. Logic models can also help to "focus and concentrate" resources against key objectives, which can be particularly beneficial and when funding is tight.

- *Work Planning*: There is no standard approach for refuge-level work planning. Many or most refuges do not have annual work plans, and among the ones that do the format and level of detail vary widely. Without a plan as to what is to be accomplished and how it will be accomplished, it is difficult to assess performance and hold managers and other staff accountable for results that should be achieved within a particular timeframe. The lack of work plans also makes it more difficult for partners to know how they can contribute to refuge needs or where in the program they may fit.
- *RAPP Performance Reporting*: As currently constructed, the RAPP system is of limited use to refuges/field units. This leaves the primary purpose as reporting program-wide accomplishments to external audiences; however, RAPP collects information on substantially more performance indicators than would seem to be required to meet external reporting needs. In addition, there is insufficient quality control in the data collection process (leading to substantial data inconsistencies and calling into question the validity of the RAPP data), and the overall process is not managed in a way that permits the system to have an analytic or management benefit to the NWRS.

Among the evaluation team's concerns with the RAPP system is that there is no strong ownership of RAPP (or performance monitoring) within the NWRS; the data are housed in multiple locations, which makes access difficult; the data itself is often difficult to interpret; errors in consistency are common; and the system is incomplete. For example, a number of indicator definitions have not been developed, and baseline measures have not been defined. Perhaps most significantly, there is virtually no analysis of the data collected. This is somewhat understandable given the issues of data quality and the system's overall complexity, but it does limit the system's utility. The NWRS would likely be better served by a somewhat simpler data monitoring system that collects less information and is more focused on providing information that can be used to analyze key aspects of NWRS' performance (i.e. collect less data, target the data collection on key issues, and analyze and use the information to understand how the NWRS is performing). The current RAPP system provides a useful foundation for building a program-level reporting system but could benefit from substantial revision and streamlining.

Budgeting: There are a number of issues with the budget process that could benefit from attention: 1) there is no standard process for budgeting within the NWRS, which makes the budget allocation process in some regions seem somewhat capricious; 2) the lack of base budgets in nearly all regions makes it more likely there will be substantial variance in budget levels from year to year and this makes it difficult for refuge manages to plan activities; and 3) because refuge budgets may shift from year to year based on regional office decisions refuge managers often do not receive their budgets until very late in the fiscal year, after the overall budgets have been received by the regional office and allocated between refuges.

Evaluation: There is relatively little formal assessment or evaluation work conducted to assess the degree that individual refuges, and refuge managers, are accomplishing their objectives. As per the above point, if individual work plans are not developed then it is difficult to assess results as management is likely to "do what was done last year" or "do what is needed" without creating any baseline or metric as to what would be a reasonable level of accomplishment within a particular period of time.

The Need to Increase Policy Consistency: This assessment has found that there is significant variance from region to region in how the Refuge System operates. Variances in practices include the lack of consistent approaches to interpreting and enforcing policy, to varying practices in terms of basic business management processes, such as work planning, staff deployment, and budgeting.

A few examples of inconsistent practices include:

- Prioritization and implementation of Big 6 recreational activities has varied between regions.
- There is inconsistent implementation of biological inventory and monitoring practices, including the use of GIS systems. The RLGIS system appears to be the GIS system most widely used by the Refuge System, but there are a number of different systems in use across the Refuge System's different regions. This means that the collection of biological information on the status of lands is not collected in compatible formats, which can make it difficult and inefficient to gather and analyze system-wide information

- There is not a consistent work planning or budgeting process among regions.
- The development and format of regional workforce plans has varied between regions.
- Refuge workstation assessments (evaluations) do not follow common protocols or guidance and are undertaken with differing levels of frequency in different regions, and in several regions are not undertaken at all.
- Law enforcement vehicles do not always meet guidelines for color and marking.
- Refuge websites use different formats, different logos and contain different types of information.
- The process of "complexing" has varied considerably between regions, with some regions embracing the concept and others not fully committing to the concept, or implementing some variation of the concept.
- The implementation of the Congressionally- mandated invasive strike force teams has varied between regions and made consistent and required reporting challenging [9]

The high degree of variance in the system can be attributed to some combination of the following: 1) the lack of clear and consistent policies and standard operating procedures within the Refuge System; and 2) the lack of central management authority or decision (enforcement) mechanisms whereby the Refuge System can ensure that Refuge System policies and practices are consistently interpreted and applied across regions.

Knowledge Management: There is not a formalized system to identify best practices, to share knowledge or to routinely bring together practitioners around common topics. The NWRS has developed many creative and highly successful programs but has not undertaken significant analysis to understand what makes these programs successful or what would be required to replicate such programs. This issue spans the range of the NWRS work—from riparian riverine restoration, to moist soils management to operating environmental education programs.

The experiences of the Refuge System are not well captured or catalogued. While there have been a significant number of studies undertaken and reports written, it can be difficult to access this information. The Refuge System does not have an effective information archive or document management system, which makes it difficult for staff to access the information that does exist.

Recommendations

Reduce administrative and reporting requirements: The Refuge System should strive to reduce administrative and reporting requirements—particularly for smaller refuges (e.g. seven or fewer staff). This will not likely be an easy task, as specific solutions are needed and many current reporting practices are required by policies generated outside of the Refuge System, e.g. DOI policies. Particular areas that may help to reduce reporting could include:

- Exempt smaller refuges from certain reporting requirements to reduce the reporting/administration burden on refuge managers and allow them to spend more time on core biology and visitor services tasks.
- Streamline RAPP reporting.
- Prioritize planning tasks and limit the number of these tasks that are emphasized in any given year.
- Revisit the necessity and utility of all annual reports, such as the energy report.
- Identify and reduce redundant data collection requirements between different systems, e.g. multiple entries for purchases and timekeeping.
- Establish a central information office to help reduce "data calls" to the field.
- Raise the accountability authority of field- based contracting managers across all regions, as has been done in Region 6

Business Management Processes

- *NWRS Strategy/ RAPP Performance Reporting*: The NWRS should revisit and adjust its strategy. There may be benefits to developing a logic model that clarifies the causal linkages that underpin the program. A causality model would help the Refuge System to clarify the cause and effect relationships betweens its principal activities and the results it hopes to accomplish and would place increased emphasis on the achievement of high-level results, such as providing satisfactory visitor services and adequate management of habitat and species. To the extent possible, habitat needs (and strategic growth decisions) should be species-driven, which will require a collaborative process with other programs within FWS (or be based on a species-driven landscape scale FWS strategy, which currently does not exist).

- *The RAPP Reporting System*: This system should be redesigned based on a clarification of its purpose. If the system is to remain primarily an external reporting tool – for reporting to FWS, DOI, OMB, and Congress— then the system should be substantially simplified to focus on areas of key interest and the number of indicators tracked should be significantly reduced (by at least 50-60%). The RAPP system would best serve the Refuge System if it were better designed to provide information that was analyzed and used to inform performance and management decision-making – it should be possible to redesign the system for this purpose while also reducing the volume of data collection (and the number of performance indicators).
- *Annual Work Planning*: A simple annual work plan template should be developed and used by all refuges. Mindful that a goal of the Refuge System should be to reduce administrative requirements placed on refuge managers, the annual work plan's format should be a relatively simple, but it should include elements that are considered standard to work planning, including assigning responsibilities, establishing a schedule and identifying the major sub-activities that are required to complete a given objective. The annual work plan should be explicitly tied to a refuge's CCP and should be used as a basis for conducting periodic refuge operational assessments. The current RAPP process helps to set annual targets but is not itself a work plan (at least as currently used by most refuges).
- *Budgeting:* A system of base budgeting should be developed for all refuges and a standard process for developing and administering base budgets should be put into place system-wide. It is suggested that this system be based on the system that is currently used in Alaska. In order to increase flexibility, consideration could be given to two-year spending authority.

Policy Development: The NWRS could benefit from a clarified policy development process. In particular, the sequence of input from various parties and the conditions that should trigger additional public review and comment opportunities should be clarified. One consideration might be to require a second round of public comment for any policy that has not been finalized within a specified period of time, for example, when a policy has not been finalized within two years of the initial public comment period.

In addition, the Refuge System could explore the feasibility of establishing an advisory council that could act as a science-based advisory group on policy development (and other initiatives). A council similar to the US Park Service's Science Advisory Council could be examined as a potential model to help ensure that NWRS policies conform to standard and best scientific practices and are consistent with the mandates of the Refuge System and the FWS. An advisory board could potentially be helpful in balancing political and science-based interests in the policy development process.

Evaluation: Periodic refuge operational assessments should be undertaken and should be a responsibility of refuge supervisors. Guidelines should be developed and implemented systemwide to ensure that some minimum number of assessments are conducted annually in each region. The evaluation process itself should draw on the involvement of staff from other refuges to provide cross-learning and should be used to identify lessons and best practices. It might also be useful to involve partners and academics in some of these assessments—or to conduct a batch of assessments that compares management practices across regions, or across a number of similar refuges (e.g. coastal wetlands), as a way to identify innovative practices.

The purpose of assessments/station reviews would be to increase performance accountability and also to generate lessons learned and best practices that could have broader application across the system. Station reviews/assessments could be conducted of refuges, regional offices and/or particular program components, e.g. wilderness management or environmental education.

Policy and Program Consistency: It is recommended that changes be implemented to enable the refuge system program to operate in a more consistent and structured manner. This should be done to promote greater program and policy consistency across regions and to increase administrative and program efficiency. While increased structure and consistency should provide several benefits to the refuge system such adjustments should not be over-done, i.e. flexibility should be maintained to continue to allow regions to prioritize issues and approaches based on local conditions.

There are a number of potential actions that could be taken to increase the refuge system's ability to increase policy consistency and standardize business practices. Many of the steps necessary to increase policy and operational consistency can be implemented by the refuge system itself under its existing operational structure. Actions that should be considered include the following:

- *Develop a better-defined set of decision- making rules and norms for use by the Regions.* For example, the Regions could agree on a decision-making model and agree that the Regional Directors or Regional Chiefs would oversee the refuge system as a stewardship board (similar to how a board of directors oversees a company). This could mean, for example, that rules are put into place as to how decisions will be made (consensus, majority, or two-thirds vote) and the group should agree to abide by its decisions. The role of the board could include policy and program direction and oversight, annual and periodic performance review, strategic planning, and budget allocation and review.

- Develop a list of practices and policies that will be considered as priorities for increasing consistency and improving performance, particularly in the areas of policy implementation and businesses processes.
- Implement a set of standard operating procedures that would apply to business management practices across all regions. In particular, practices could be standardized in the areas of budgeting, work planning, and station assessments/evaluation.
- Annually prioritize programmatic areas that can be reviewed and improved based on analysis, evaluations and the identification of best practices.
- When providing outreach and orientation services on refuge policy the refuge system should be mindful to include the Service Directorate and their staff. This may help to build a common base of knowledge and familiarity on refuge policies across the FWS' leadership cadre.

These above measures should help to improve policy and operational consistency within the refuge system.

While the above steps will help to improve consistency, these steps principally rely on voluntary cooperation and goodwill and stop short of identifying a clear policy enforcement authority or mechanism for ensuring policy implementation consistency when there are professional differences of views between FWS senior decision makers.

The MSI evaluation team noted that the issue of refuge system's structure and authority has been previously raised by several well-respected study teams. Such studies include the 1968 Leopold Report and the 1992 Beattie Report, both of which recommended the refuge system be managed under a centralized line authority structure. This assessment did not study the issue of line management authority and therefore does not make a conclusion or recommendation as to whether or not such a structure would meaningfully improve the refuge system's performance or consistency. The issue is noted for the following reasons: it has been commented on by prior studies; it was a topic raised in MSI interviews by a large number of refuge managers and senior staff, and; this evaluation concluded that inconsistent policies and practices are a hindrance to improving the refuge system's overall effectiveness. As noted earlier, however, the refuge system does have the authority and ability to implement meaningful improvements in this area under its existing structure of operations.

Knowledge Management: The NWRS should implement a Knowledge Management System. The goals of the system should include:

- Develop an electronic library of key documents that would be easily accessible to all staff;
- Conduct program-level and topical studies to identify best practices and to then publish and identify such practices;
- Incorporate best practices into training materials, guidance, reviews, and standard practices so as to raise the quality level of implementation practices across the Refuge System; and
- Establish topical learning communities and provide practitioner communities the resources required to develop and analyze strategies and approaches and to be able to get together periodically to share information.

Consideration should be given to creating a Knowledge Management Unit, which would be responsible for program reporting (RAPP), evaluation, generating, and disseminating lessons learned and best practices, archiving documents and responding to external information requests (together with public relations staff)

ADDENDUM: KNOWLEDGE

The field of Knowledge Management has developed over the past ten years or so in response to the rapidly increasing body of knowledge available in the world and to the improvements in technology that make it possible to easily collect and share knowledge across geographic boundaries and professional disciplines. Knowledge Management comprises a range of practices used by organizations to identify, create, represent, and distribute knowledge for reuse, awareness and learning. It has been an established discipline since 1995 with a body of university courses and both professional and academic journals dedicated to it. Most large companies have resources dedicated to Knowledge Management, often as a part of 'Information Technology' or 'Human Resource Management' departments, and sometimes reporting directly to the head of the organization. Knowledge Management programs are typically tied to organizational objectives and are intended to achieve specific outcomes, such as shared intelligence, improved performance, competitive advantage, or higher levels of shared innovation. Observations on this topic from our assessment include:

- It was difficult to obtain historical documents on the performance of the NWRS or on particular programs within the NWRS – there does not appear to be an electronic library or database of documents, or an easy way to know what documents exist. Key documents of significance to this evaluation were found only through happenstance or serendipity as, for example, when an evaluation of the CCP process was "discovered" accidentally in the course of a site visit.
- There is a lack of guidance in terms of how to implement specific programs. For example, there does not appear to be much guidance for structuring interpretive or environmental education programs, including what types of activities are most effective for which age groups. Another example is the Refuge System's experience with "complexing," which differs from region to region without a clear understanding of the differences in approaches or the benefits of one approach over another.

> *Knowledge Management (KM)* is the process through which organizations generate value from their intellectual and knowledge-based assets. Most often, generating value from such assets involves codifying what employees, partners and customers know, and sharing that information among employees, departments and even with other companies in an effort to devise best practices. It's important to note that the definition says nothing about technology; while KM is often facilitated by IT, technology by itself is not KM.

There are exceptions to the above observations, such as the process used to analyze the effectiveness of CCPs and to incorporate what is learned into updated training courses. And it should also be noted that program implementation guidance and best practices are included in many of the refuge management courses provided by the National Conservation Training Center. In general, however, the NWRS operates with a great deal of autonomy from one region to the next and there has not been sufficient analysis of many of the system's most important areas of emphasis, such as comparative studies on biological monitoring practices.

Without a more cohesive approach to analyzing program effectiveness, identifying best practices and disseminating this information the Refuge System is not able to take advantage of the opportunities offered by a "systems" approach to management and too often results and learning are limited by the skills and experience of individuals. While these individuals may be highly skilled they often work in isolation from other specialists working on similar issues. Because the Refuge System is so decentralized, there are a large number of examples within the system of highly creative and effective programs – such as the environmental education program at Fergus Falls, or the fire management program at Okefenokee – but the lessons from these programs are not well known throughout the Refuge System. There are a number of advantages the Refuge System could gain from adopting a Knowledge.

Management approach to how it conducts business. Basic opportunities include:

- Analyzing program approaches to gain increased effectiveness and cost efficiency: For example, the NWRS has substantial experience with "complexing", which has been implemented differently in different regions. This experience could be reviewed to determine if there are guiding principles that can be applied across the system.
- Reduce the development of redundant systems. For example, it is not necessary that all regions develop their own work planning or budgeting processes, or that all regions have their own process for evaluating refuge performance, or that they all develop their own environmental education materials. Common systems and approaches could be developed and then applied throughout the system.
- Develop tool kits for use by refuge managers across the system. For example a toolkit/backpack could be developed for use by small refuges in offering environmental education opportunities to school kids. The kits could be adapted for use in different habitats but could be general enough to allow programs to be offered at many of the smaller refuges that do not have the staff to develop such materials.

Best practices should be identified by analyzing similar and contrasting experiences and identifying factors the lead to increased effectiveness. As necessary, best practice approaches should be supported through document development, networking groups, periodic professional meetings of practitioners, development and dissemination of tool kits and establishment of support centers that can provide supporting material and expertise.

Within all management systems there is a need to balance flexibility with structure. Flexibility enables managers to be responsive to local circumstances and to develop creative systems to address issues, whereas structure takes advantage of opportunities to ensure that best practices are identified and adopted throughout the system, provides structured opportunities for professionals to learn from one another, and seeks opportunities to gain efficiencies through standardization and establishment of central support structures. The Refuge System is replete with examples of innovation and creativity, but does not have systems in place to ensure that best practice ideas are well-known or can be widely adopted.

REFERENCES

[1] National Wildlife Refuge System Improvement Act of 1997; Public Law 105-57.

[2] Except for two surveys (partner survey and state fish and game agency survey), evaluation activity was completed by September 2007; the surveys were completed in April and May 2008.

[3] "Managing Biological Integrity, Diversity, and Environmental Health in the National Wildlife Refuges: An Introduction to the Symposium," Robert Fischman and Vicky Meretsky, Natural Resources Journal, University of New Mexico School of Law 44(2004): page 940.

[4] Note: The Partners survey is used in this report to provide supplemental data to other findings. A representative sample of NWRS partners could not be developed because there was not adequate information on the current total universe of NWRS partners. As a result, the data from this survey is not generalizable to the overall population of partners. However, the information provides value by providing additional perspective on the Refuge System's performance and is considered, along with other data, in the analysis of the Refuge System's effectiveness

[5] During a workshop held after the conclusion of field level data collection by the evaluation team, Refuge System staff from both Washington and the field noted that state and county departments of transportation often pose a substantial constraint to the placement of directional signage on non- refuge roads.

[6] The Refuge System should review the format and content of the websites of other USG land management agencies to provide ideas and possible options for any revisions to the Refuge System websites. The website of the National Park Service – and the websites of each National Park – is particularly instructive. That is, the Park Service website format is exactly

consistent for all Park Service units, but allows flexibility for content specific to each Park. Use of a consistent format has two obvious advantages: (1) it presents to the public a single institutional identity, and (2) it greatly increases the ease with which a user can navigate any individual website within the system and quickly locate the information they are interested in

[7] Many national parks offer "discovery backpacks" for use by visiting school children. The backpacks may contain binoculars, bird lists, and guides on local ecology, such as wetland areas or upland forests. The topical material included in the backpacks varies from park to park and is selected from available national resources, based on the ecology or geography of the area

[8] Note: The data in the chart is taken from the 2005 Annual Volunteer Report in order to have comparable data over a longer time period. RAPP data on page 2 varies from the information in the Annual Volunteer Report

[9] Several of these concerns were raised in a workshop with Refuge System managers, including the lack of consistency in the procurement of law enforcement vehicles and the lack of adherence to requirements in the work of the invasive species strike force teams.

INDEX

A

B

C

D

E

F

J

L

M

N

O

P

Q

R

U

V

W